2

MARINE BIOLOGY

MARINE SEDIMENTS

FORMATION, DISTRIBUTION AND ENVIRONMENTAL IMPACTS

MARINE BIOLOGY

Additional books in this series can be found on Nova's website under the Series tab.

Additional e-books in this series can be found on Nova's website under the eBooks tab.

MARINE BIOLOGY

MARINE SEDIMENTS

FORMATION, DISTRIBUTION AND ENVIRONMENTAL IMPACTS

SHIRLEY WILLIAMS
EDITOR

New York

Library of Congress Cataloging-in-Publication Data

Names: Williams, Shirley (Editor at Nova Science Publishers), editor.
Title: Marine sediments : formation, distribution and environmental impacts / editor, Shirley Williams.
Description: Hauppauge, N.Y. : Nova Science Publisher's, Inc., 2016. | Includes bibliographical references and index.
Identifiers: LCCN 2016014498 (print) | LCCN 2016021990 (ebook) | ISBN 9781634851275 (softcover) | ISBN 9781634851473 (Ebook) | ISBN 9781634851473 ()
Subjects: LCSH: Marine sediments--Analysis.
Classification: LCC GC380.2.A52 M37 2016 (print) | LCC GC380.2.A52 (ebook) | DDC 551.46/86--dc23
LC record available at https://lccn.loc.gov/2016014498

Published by Nova Science Publishers, Inc. † New York

CONTENTS

PREFACE

This book discusses the formation, distribution and environmental impacts of marine sediments. The first chapter describes the isolation, taxonomic approach, diversity, secondary metabolites and activities of actinobacteria. Chapter Two aims at the verification of anisotropy of magnetic susceptibility (AMS) usability as a textural indicator of rocks, with special emphasis on origin of fluid pathway within tight turbidite sandstones burying a foreland basin. The final chapter reviews aluminum impact on the growth of benthic diatom.

Chapter 1 – The widely distributed component of marine sediments in many areas such as in USA, Mexico, Bahamas, Papu New Guinea, the Republic of Palau, Guam, Fiji, Chile, Antarctica, Norway Mediterranean, Australia, India, China, Republic of Korea, Japan and Thailand, associated microbial communities represented different taxonomic groups at the phylum level including Acidobacteria, Planctomycetes, Actinobacteria, Gamma-, Alpha -and Delta-proteobacteria based on the 16S rRNA gene sequences and phylogenetic analyses. This divergence was mainly triggered by the dominance of Acidobacteria and Actinobacteria. A majority of Actinobacteria in sediment samples contained predominantly the genera *Streptomyces*, *Micromonospora*, *Actinocorallia*, *Actinomadura*, *Knoellia*, *Glycomyces*, *Nocardia*, *Nocardiopsis*, *Nonomuraea*, *Pseudonocardia*, *Rhodococcus*, *Saccharomonospora*, *Streptosporangium*, *Salinispora*, and *Sciscionella.* Actinomycetes in marine sediments produced a wide variety of secondary metabolites with diverse biological activities. *Streptomyces* strains produced various anti-cancer and anti-tumor bioactive compounds such as actinoranone, chlorizidine, galvaquinones, glucopiericidin, grincamycins, lobophorins, marfuraquinocins, nitropyrrolins, spiroindimicins, strepnonesides andsungsanpin while the rare actinomycetes in genera *Actinoalloteichus*,

Actinomadura, *Marinactinospora*, *Marinispora*, *Micromonospora*, *Nocardiopsis*, *Saccharomonospora*, *Salinispora*, and *Verrucosispora* produced cyanogrisides, fijiolides, fluostatins, marthiapeptide, marinactinones and nocardiopsins. At present, bioactive metabolite of obligate marine actinomycete, *Salinispora tropica*, salinosporamide A is a potent proteasome inhibitor being used as the anticancer agent that currently being evaluated in phase I clinical trials as monotherapy. Additionally, many compounds firstly reported from marine sediment actinomycetes may be guides for drug discovery, for example, abyssomicin C from *Verrucosispora*, showed *in vivo* activity against *Mycobacterium tuberculosis*, and marinomycin A from *Marinispora* is potent antitumor agent with substantial activities against selected human tumor cells and drug-resistant pathogenic bacteria. In this chapter, the isolation, taxonomic approach, diversity, secondary metabolites and activities of actinobacteria are described.

Chapter 2 – Magnetic techniques that use anisotropy of magnetic susceptibility (AMS) act as a proxy of preferred permeable orientation in basin-filling sediments, when it is applied on samples impregnated with a magnetic suspension. The unique method for quantifying heterogeneity in rocks is reviewed and its value for reconstruction of the preferred direction of pore fluid flow is reassessed critically. The authors also present results of their experiments, which dealt with secondary fracture networks developed in tight sandstones burying a foreland basin on an arc-arc collision zone. Directional analysis of AMS ellipsoid implies tectonic control on rupture development under strong trans-compressive regime. Micro-focus three-dimensional density imaging of test pieces has shown a substantial variation in pore fabric reflecting inhomogeneous impregnation of magnetic fluid within rocks.

Chapter 3 – A laboratory experiment was performed, in tidal artificial conditions, with natural defaunated sediments that were artificially contaminated with aluminium salts. Benthic diatoms were cultivated on these contaminated sediments in a tidal mesocosm with natural seawater, in order to assess the aluminium impact on the growth of benthic diatom. The sediments were also leached by HCl (1M) to estimate the available/mobile part of the aluminium in the sediment. This experiment highlighted that Al-contamination sediments had a strong effect on diatom communities growing at sediments surface. Drastic negative effect is perceptible beyond 10 mg of aluminium salts added by kg of natural sediment (which is a low contamination stress with regard to the Al initial content). The 1M HCl-extraction, widely used in the literature for metals mobility evaluation in marine sediments seems to underestimate aluminium availabilities for phytobenthic communities.

In: Marine Sediments
Editor: Shirley Williams

ISBN: 978-1-63485-127-5

Chapter 1

ACTINOBACTERIA FROM MARINE SEDIMENTS: DIVERSITY AND SECONDARY METABOLITES

Khomsan Supong and Somboon Tanasupawat*
Department of Biochemistry and Microbiology,
Faculty of Pharmaceutical Sciences,
ChulalongkornUniversity, Bangkok, Thailand

ABSTRACT

The widely distributed component of marine sediments in many areas such as in USA, Mexico, Bahamas, Papu New Guinea, the Republic of Palau, Guam, Fiji, Chile, Antarctica, Norway Mediterranean, Australia, India, China, Republic of Korea, Japan and Thailand, associated microbial communities represented different taxonomic groups at the phylum level including Acidobacteria, Planctomycetes, Actinobacteria, Gamma-, Alpha- and Delta-proteobacteria based on the 16S rRNA gene sequences and phylogenetic analyses. This divergence was mainly triggered by the dominance of Acidobacteria and Actinobacteria. A majority of Actinobacteria in sediment samples contained predominantly the genera *Streptomyces*, *Micromonospora*, *Actinocorallia*, *Actinomadura*, *Knoellia*, *Glycomyces*, *Nocardia*, *Nocardiopsis*, *Nonomuraea*, *Pseudonocardia*, *Rhodococcus*,

* Corresponding author: SomboonTanasupawat .Tel +66-2-2188376, Fax +66-2-2545195, E-mail :Somboon.T@chula.ac.th.

Saccharomonospora, *Streptosporangium*, *Salinispora*, and *Sciscionella*. Actinomycetes in marine sediments produced a wide variety of secondary metabolites with diverse biological activities. *Streptomyces* strains produced various anti-cancer and anti-tumor bioactive compounds such as actinoranone, chlorizidine, galvaquinones, glucopiericidin, grincamycins, lobophorins, marfuraquinocins, nitropyrrolins, spiroindimicins, strepnonesides and sungsanpin while the rare actinomycetes in genera *Actinoalloteichus*, *Actinomadura*, *Marinactinospora*, *Marinispora*, *Micromonospora*, *Nocardiopsis*, *Saccharomonospora*, *Salinispora*, and *Verrucosispora* produced cyanogrisides, fijiolides, fluostatins, marthiapeptide, marinactinones and nocardiopsins. At present, bioactive metabolite of obligate marine actinomycete, *Salinispora tropica*, salinosporamide A is a potent proteasome inhibitor being used as the anticancer agent that currently being evaluated in phase I clinical trials as monotherapy. Additionally, many compounds firstly reported from marine sediment actinomycetes may be guides for drug discovery, for example, abyssomicin C from *Verrucosispora*, showed *in vivo* activity against *Mycobacterium tuberculosis*, and marinomycin A from *Marinispora* is potent antitumor agent with substantial activities against selected human tumor cells and drug-resistant pathogenic bacteria. In this chapter, the isolation, taxonomic approach, diversity, secondary metabolites and biological activities of actinobacteria are described.

1. INTRODUCTION

The component of marine sediments associated microbial communities represented different taxonomic groups mainly by the dominance of Acidobacteria and Actinobacteria (Polymenakou et al., 2009). The class *Actinobacteria* Stackebrandt et al. 1997 (Gram- positive bacteria with a high G + C content) are a well-known source of secondary metabolites and the marine counterparts are proving to be a richsource of novel bioactive compounds (Bérdy, 2005; Fiedler et al. 2005; Blunt et al. 2007). The diversity of actinobacteria in marine habitats is little known based on the description of the first sea-water dependent actinobacterial genus *Salinispora* (Maldonado et al. 2005a) that supports the view of marine ecosystem, an untapped source of microorganisms of economical importance. It is only more recently that marine-derived actinomycetes have become recognized as a source of novel antibiotics and anti-cancer agents with unusual structures and properties (Bredholt et al., 2008). Some of the unusual structures and properties of compounds isolated from marine sources and the fact that 58% of the isolated actinomycetes from marine sediments collected around Guam in the Pacific ocean required sea water for growth (Jensen et al., 2005) implies that one may find microorganisms adapted to the marine environment and producing compounds not found among microorganisms adapted to the terrestrial sources. Actinomycetes are best known as soil bacteria and were generally believed to occur in the ocean largely as dormant spores that were washed into the sea (Goodfellow and Haynes, 1984).

The distributions and ecological roles of actinomycetes in the marine environment, and the extent to which obligate marine species occur, have remained an unresolved issue in marine microbiology. Recently, Jensen et al. (2005) reported the cultivation from marine sediments of a major new group of marine actinomycetes (originally called MAR1, *Micromonosporaceae*) for which the generic epithet '*Salinospora*' was proposed (Mincer et al., 2002) and later was revised to be *Salinispora* gen. nov., which comprised only three species *Salinispora aenicola*, *Salinispora tropica* and *Salinispora pacifica* (Maldonado et al., 2005a; Ahmed et al., 2013). All *Salinispora* strains required seawater and more specifically, sodium for growth indicating a high level of marine adaptation. This new taxon is the source of novel secondary metabolites including salinosporamide A, a potent anticancer agent specifically targeting the 20S subunit of the mammalian proteasome (Feling et al., 2003). The discovery of this taxon provides clear evidence for the existence of autochthonous populations of marine actinomycetes, and the compounds being discovered from them indicate that cultured strains are an important resource for novel secondary metabolites. In addition, the taxon has proven to be a productive source of structurally unique and biologically active secondary metabolites (Feling et al., 2003; Jensen et al., 2005). To enhance the chance of recovering the novel actinobacteria from the sea by improving the selective isolation trategies, sample pretreatments, selective isolation media and the dereplication of isolate collections can still help us to recover putative novel species from one of the most important biotechnologically microbial groups. Thus, there is evidence that marine actinobacteria represent an autochthonous yet little understood component of the sediment microbial community as well as a useful resource for pharmaceutical discovery.

2. Samples Collection, Isolation and Cultivation of Actinobacteria

2.1. Samples Collection

Samples for the selective isolation procedures were taken from various places such as the samples were obtained from the first 1 to 5 cm of sediment by scuba collection at depths of 10 to 40 m, collected at the Bismarck Sea, in Madang Province, the northern coast of Papua New Guinea, and to the Solomon Sea, in Milne Bay Province, and placed in sterile 50-ml conical tubes and the samples were kept at room temperature during the expedition and at 4°C upon return to the laboratory (Magarvey et al., 2004) while the marine sediments from a depth of 300 m was kept under liquid nitrogen until processing (Maldonado et al. 2009). Each wet sediment sample (approximately 50 mg) was used to inoculate on NaST21Cx agar plate supplemented with 25 μg of cycloheximide/ml. (Magarvey et al., 2004) as shown in Table 1. Whatman no.1 sterile filter paper disks (precut to fit the agar surface) were placed on the agar, and the sediment material was dispersed evenly on the surfaces of the cellulose disks. The plate was incubated in a humidified chamber at 30°C for 30 to 90 days. Following incubation, selected colonies were streaked on ISP-2 medium (BD Diagnostics, Sparks, Md.) prepared with artificial seawater (ISP-2/ASW) and containing cycloheximide (25 μg/ml) and nalidixic acid (25 μg/ml). Mincer et al. (2002) collected the top 1 cm of sediment samples in sterile 50-ml plastic Whirl-Pak bags (NASCO, Modesto, Calif.) by divers using scuba gear. Sediment sample depth ranges, numbers, locations, and dates were as follows: 0 to 30 m, Sea of Cortez, Mexico. The samples were processed as soon as possible after collection (generally, 4 h) using the selective methods of (i) stamping and (ii) dilution and heat shock or both (as described below) and were inoculated onto isolation media (M1 to M5). The dilution-and-heat-shock method was carried out as follows: 1 ml of wet sediment was added to 4 ml of sterile seawater, heated for 6 min at 55°C, vigorously shaken, and further diluted (1: 4) in sterile seawater, and 50 l of each dilution was inoculated by spreading with a sterile glass rod onto agar-based isolation media. The stamping method was carried out as follows: 10 ml of wet sediment was aseptically placed into a sterile aluminum dish, dried (ca. 24 h) in a laminar flow hood, ground lightly with a pestle, pressed into a sterile foam plug (14 mm in diameter), and inoculated onto agar media by stamping eight or nine times in a circular fashion, giving a serial dilution effect. All media were prepared with 100% filtered natural seawater. Bredholdt et al., (2007) collected the marine sediment samples by scuba divers close to shore outside the Trondheim Biological station, Norway, 63′ 27 ‘N and 10′ 21′-E). Only the upper 5 cm of the sediments were collected. The samples including, sample 2: Depth 28 m, fine mud and clay, no vegetation, sample 5: Depth 27 m, fine mud and sand, no vegetation, sample 7: depth 6 m, clay and stones, vegetation, brown layer from sedimented algae and sample 10: depth 4.5 m, fine mud with small stones, upper layer covered with diatom and bacteria, some vegetation. Sediments from 4.5, 6.0, 27 and 28 m depths were collected by scuba divers, while sediments from 450 m depth were sampled with a box-corer. The upper 5 cm of the sediments were collected in zip-lock bags (scuba) or with a sterile spade (box–corer) and transferred to 1 liter sterile plastic containers (Bredholdt et al., 2008).

2.2. Selective Isolation Procedure

The wet sample from marine sediment (1 g) of each Gulf, was transferred to a 15 ml centrifuge tube which contained 9 ml of seawater solution (3.5%; Instant Ocean, USA) and 1 g of sterile glass beads (710 – 1.180 lm; Sigma, USA). Each centrifuge tube was then shaken for 1 h in a blood tube rotator model SB2 at fixed speed (Fischer Scientific, USA). A second and third series of dilutions (1 ml into 9 ml of sea water) were then prepared for each sample (10^{-2} and 10^{-3}). The three dilutions were then employed to inoculate (100 μl) a set of selective isolation plates by duplicate (Maldonado et al. 2009). Bredholdt et al. (2007) prepared the samples by aseptically diluting the wet sediment (1 ml each in 9 ml sterile water). The actinobacteria selective treatments were performed on dried sediments (Speed vac. 30°C, 16 h) including dry heat (120°C, 60 min), phenol (1.5%, 30 min at 30°C), dry heat and phenol, dry heat and benzethonium chloride (0.02%, 30 min at 30°C), as well as pollen baiting as previous reported (Hayakawa et al., 1991; Palleroni, 1980). Ten microlitres of the dilution were spread over the surface of isolation media in Petri dishes. Isolation plates were incubated at 20 28C for 2 – 6 weeks (Bredholdt et al. 2008; Maldonado et al. 2009). Actinomycetes generally appeared after 2 to 6 weeks of incubation and were considered to be any colony with a tough leathery texture, dry or folded appearance, and branching filaments with or without aerial mycelia. They were detected by eye and by using a binocular microscope (Olympus Optical Co., Ltd, Tokyo, Japan) fitted with a long working distance objective.

Table 1. The isolation media for actinomycetes from marine sediments

Medium	Medium components	References
M3	K_2HPO_4, 0.466 g; Na_2HPO_4, 0.732 g; KNO_3, 0.1 g; NaCl, 0.29 g; $MgSO_4$:$7H_2O$, 0.10 g; $CaCO_3$, 0.02 g; Na-propionate, 0.20 g; $FeSO_4$:$7H_2O$, 200 µg; $ZnSO_4$:$7H_2O$, 180 µg; $MnSO_4$:$4H_2O$, 20 µg; thiamine-HCl, 4.0 mg; agar, 18 g in 1 l of natural seawater	Maldonado et al., 2005
M4	Chitin, 2 g; agar 18 g in 1 l of natural seawater	Mincer et al., 2002
M5	Agar, 18 g in 1 l of natural seawater	Mincer et al., 2002
Medium 1 (AMM)	Agar, 8 g; starch, 10 g; yeast extract, 4 g; peptone, 2 g in 1 l of natural seawater	Jensen et al., 2005
Medium 2 (NPS)	Noble agar, 8 g; NPS (nutrient poor sediment) extract, 100 ml in 1 l of natural seawater. NPS extract was prepared by washing (extracting) 900 ml (wet volume) of sand collected from a high-energy beach with 500 ml of sea water	Jensen et al., 2005
Medium 3 (NRS)	Noble agar, 8 g; NRS (nutrient rich sediment) extract, 100 ml in 1 l of natural seawater. NRS extract was prepared as above using 300 ml (wet wolume) of sediment collected at low tide from a mangrove channel	Jensen et al., 2005
Medium 5 (SMC)	Noble agar, 8 g; mannitol, 500 mg; casamino acids, 100 mg; nystatin (50 µg/ml) in 1 l of natural seawater	Jensen et al., 2005
Medium 4 (SHG)	Noble agar, 8 g; humic acid sodium salt, 100 mg; galactose, 500 mg; nystatin (50 µg/ml)	Jensen et al., 2005
Medium 6 (SMP)	Noble agar, 8 g; mannitol, 500 mg; peptone 100 mg; rifampicin (5 µg/ml) in 1 l of natural seawater	Jensen et al., 2005
Medium 7 (SNC)	Agar, 8 g; novobiocin (25µg/ml) in 1 l of natural seawater	Jensen et al., 2005
Medium 8 (SPC)	Agar, 8 g; polymixin B sulfate (5 µg/ml) in 1 l of natural seawater	Jensen et al., 2005
Medium 9 (SRC)	Agar, 8 g; rifampicin (5 µg/ml) in 1 l of natural seawater	Jensen et al., 2005
NaST21Cx agar	Solution A (750 ml of artificial seawater containing K_2HPO_4, 1.0 g ;and Bacto agar, 10 g) and solution B (250 ml of artificial seawater containing KNO_3, 1.0 g; $MgSO_4$, 1.0 g; $CaCl_2$.$2H_2O$, 1.0 g; $FeCl_3$, 0.2 g and $MnSO_4$:$7H_2O$, 0.1 g). Solutions A and B are autoclaved separately and mixed, supplemented with 1 ml of trace element solution	Magarvey et al., 2004
Raffinose histidine agar (RH)	Raffinose, 10 g; *L*-histidine, 1.0 g; K_2HPO_4, 1.0 g; $MgSO_4$:$5H_2O$, 0.5 g; $FeSO_4$, 0.01 g; agar, 15 g in 1 l of natural seawater	Maldonodo et al., 2009
Soil agar-sea salt	Agar 20 g; sea salt, 30 g in 1 l of soil extract solution. 1 ml of vitamin complex solution (calcium pantothenate, 10.0 mg; nicotinic acid, 10.0 mg; thiamine-HCl, 1.0 mg; biotin, 1.0 mg in 20 ml of tap water. Soil extract was prepared by mixing 200 g soil in 1 l water, followed by boiling the mixture for 30 min.	Bredholdt et al., 2007

2.3. Media for Selective Isolation

Mincer et al. (2002) isolated actinomycetes using various media (M1 to M5): M1, 10 g starch, 4 g yeast extract, 2 g peptone, 18 g agar, and 1 liter of natural seawater; M2, 6 ml glycerol, 1 g arginine, 1 g K_2HPO_4, 0.5 g $MgSO_4$, 18 g agar, and 1 liter of natural seawater; M3, 6 g glucose, 2 g chitin (United States Biochemical, Cleveland, Ohio), 18 g agar, and 1 liter of natural seawater; M4 2 g chitin, 18 g agar, and 1 liter of natural seawater; and M5, 18 g agar and 1 liter of natural seawater. All isolation media were amended with filtered (0.2-m pore size) cycloheximide (100 g/ml) and rifampin (5 g/ml), after autoclaving. Mincer et al. (2005) collected marine sediment samples around the islands of the Bahamas. Sediment samples ranged from fine carbonate muds to coral rubble and were collected using SCUBA or a modified, surface deployed sampler. All sediment samples were processed in the field as soon as possible after collection by using desiccation and heat shock as selective cultivation methods. These methods were designed to reduce the numbers of Gram-negative bacteria and to enrich for slow-growing, spore-forming actinomycetes. Treated samples were then inoculated onto medium M1 or M5 (Table 1) and incubated for 4 to 8 weeks at room temperature. Novobiocin (10 μg/ml) or rifampin (5 μ/ml) was added to reduce the number of unicellular bacteria. The antifungal agent cycloheximide (100 μg/ml) was added to all isolation media. Enrichment cultures were prepared in 20-ml vials by adding 1.5 g of wet sediment (homogenized and previously frozen) to either 10 ml of seawater enriched with crude chitin (0.1% w/v), 10 ml of sediment extract [SE; autoclaved supernatant from a 0.5% (w/v) sediment-seawater solution], or 10 ml of medium M1 low (0.2% starch, 0.08% yeast extract, 0.04% peptone, seawater). Each of the three enrichment conditions was supplemented with either 5 μg of rifampin/ml or 25 μg of novobiocin/ml (final concentrations). Similar enrichments were prepared in the field with 0.5 g sediment and one of the following antibiotics (final concentrations): kanamycin (20 μg/ml), novobiocin (10 μg/ml), vancomycin (5 μg/ml), gentamicin (2 μg/ml), or tetracycline (4 μg/ml). All enrichment cultures were incubated for 4 to 15 weeks at room temperature and observed at x10 to x 64 magnification using a stereomicroscope. Bacteria that formed visible mycelia were harvested directly from the enrichment cultures by using a sterile pipette, serially washed three times in 10 ml of sterile seawater, and plated on medium M5.

Jensen et al. (2005) collected shallow sediments from the depths of 1–20 m, Guam Island while the remaining sediments were collected using a modified, surface-deployed sediment samplerto depths of 570 m. All samples were processed within a few hours of collection using a variety of techniques designed to reduce the numbers of Gram-negative bacteria and to enrich for slow-growing, spore-forming actinomycetes. Samples were processed andinoculated onto various agar media [medium 1 to medium 12, [Medium 1 (AMM), Medium 2 (NPS), Medium 3 (NRS), Medium 4 (SHG), Medium 5 (SMC), Medium 6 (SMP), Medium 7 (SNC), Medium 8 (SPC), Medium 9 (SRC), Medium 10 (SSC), Medium 11 (STC), Medium 12 (SMY)] (Table 1) using one, or in some cases (especially for the deeper sediments) as many as three, of the eight methods described below. All algal samples were processed using method 1 (with grinding) while all sponges were processed using method 3. Method 1 (dry/stamp). Sediment was dried overnight in alaminar flow hood and, when clumping occurred, groundlightly with an alcohol-sterilized mortar and pestle. An autoclavedfoam plug (2 cm in diameter) was pressed onto thesediment and then repeatedly onto the surface of an agarplate in a clockwise direction creating a serial dilution effect. Method 2 (dry/scrape). This method was used for small rocks that had been dried overnight in a laminar flow hoodand then scraped with a sterile spatula generating a powderthat was processed as per method 1. In some cases, thepowder was collected with a wet cotton-tipped applicator or the rock was rubbed directly with the applicator which wasthen used to inoculate the surface of an agar plate. Method 3 (dry/dilute). Dried sediment (*c.* 0.5 g) was dilutedwith 5 ml of sterile (autoclaved) seawater (SSW). The diluted sample was vortex mixed, allowed to settle for a few minutes, and 50 ml of the resulting solution inoculated onto the surfaceof an agar plate and spread with an alcohol-sterilized glassrod. Method 4 (dilute/heat). Dried sediment was volumetricallyadded to 3 ml of SSW (dilutions 1:3 or 1:6), heated to 55°C for 6 min, and 50 – 75 ml of the resulting suspension inoculatedonto an agar plate as per method 3. Method 5 (dilute/heat/2). Dried sediment was treated as permethod 4 (dilution 1:6) with the addition of a second heattreatment at 60°C for 10 min. Method 6 (dry/stamp + dilute/heat). The surface of an agar medium was inoculated using a sample treated as permethod 1. The dried sediment was then processed usingmethod 4 and the same agar plate inoculated a second time with the heat-treated samples. Method 7 (freeze/dilute). Wet sediment was frozen at -20°C for at least 24 h, thawed, volumetrically diluted in SSW (1:3–1:120 depending on particle size), and 50 ml of the resulting suspension inoculated onto the surface of an agar plate asper method 3. Method 8 (freeze/dilute/2). Wet sediment was treated as per method 7 except that the thawed and diluted sample wasincubated at room temperature for 48 h before inoculation onto the surface of an agar plate. Processed samples were inoculated as described aboveonto the surface of from one to eight of the following agarmedia. All media were prepared with 1 l of natural seawater and contained the anti-fungal agents cycloheximide (100 mg/ml) and, when listed, nystatin (50 mg/ml). Inoculated Petri dishes were incubated at room temperature (*c.* 28°C) and monitored periodically over 3 months for actinomycete growth. Actinomycetes were quantified on each plate by eye and with the aid of a Leica MZ6 stereomicroscope (x10 –x64).

Bredholdt et al. (2007) isolated actinobacteria using the following media: *soil agar*: 1 l of filtered soil extract, 20 g agar, pH 7.0 (soil extract was prepared by mixing 200 g soil in 1 l water, followed by boiling the mixture for 30 min); 1 ml of vitamin complex solution (10.0 mg of calcium pantothenate, 10.0 mg of nicotinic acid, 1.0 mg of thiamin chloride, 1.0 mg of biotin, 20 ml of tap water) was added into soil agar. *Soil agar* with 3% sea salt added. *Modified organic agar 2 Gause*: 5.0 g of peptone, 3.0 g of tryptone, 10 g of glucose, 5.0 g of NaCl, 1 l of tap water. *Modified organic agar 2 Gause* with 3% sea salt added. *Modified organic agar 2 Gause* supplemented with tobramycin (10 mg/ml) (Terekhova et al., 1991). *Modified organic agar 2* Gause supplemented with rubomycin (5 mg/ml) (Lavrova et al., 1972). *Mineral agar 1 Gause*: 20.0 g of starch-soluble, 0.5 g of K_2HPO_4, 0.5 g of $MgSO_4$, 1.0 g of KNO_3, 0.5 g of NaCl, 0.01 g of FeSO4, 20.0 g of agar, 1 l of tap water. *Soy-bean meal agar*: 3.0 g of soy-bean meal, 0.2 g of KNO_3, 0.5 g of K_2HPO_4, 0.4 g of $MgSO_4$, 20.0 g of agar, 1 l of distilledwater. *Soy-bean meal agar* with 3% sea salt added. *Yeast-corn-starch agar*: 10.0 g of starch-soluble, 10 g of yeastextract, 10.0 g of corn meal extract, 2.0 g NaCl, 20 g of agar, 1 l of water. *Pea-meal agar*: 10.0 g of pea meal, 10.0 g of glucose, 5.0 g of NaCl, 1.0 g of $CaCO_3$, 20 g of agar, 1.0 l of tap water. The pH of all media was adjusted to 7.0 – 7.5. Nalidixic acid (10 mg/ml) and nystatin (50 mg/ml) were supplemented to all media to inhibit the growth of Gram-negative bacteria and fungi. For isolation, a variety of methods was used, including the pretreatment of sediment samples with physical factors: UV irradiation (Galatenko and Terekhova, 1990), super high frequency (SHF) radiation (Bulina et al., 1997, 1998), extremely high frequency (EHF) radiation (Li et al., 2002), cold-shock by freezing sediment samplesat -18°C. UV-irradiation of the wet sediment suspension (5 ml) was performed in open Petri dishes for 30 s. with the use of UVlamp at emission wavelength of 254 nm and power of 15 W. The distance from the irradiation source was 20 cm. SHF radiation treatment of the suspension (2.5 ml) placed intosterile Eppendorf tubes was carried out in a microwave ovenat a frequency of 2460 MHz and power of 80

W for 45 s. EHF-radiation treatment of the suspension (5 ml) was carriedout in Petri dishes from the bottom. Emitted radiation had a non-thermal intensity and was amplitude-modulated at a frequency of 1 kHz within wavelength band of 8 – 11.5 mm using industrial generator (Russia). In addition, the two-layer soil agar with the following transfer of the upper layer to organic agar 2 Gause was used (Galatenko et al., 2004). Before inoculation agar media in Petri dishes were dried for 30 min to prevent water condensation on agar surface. The isolation plates were incubated at 28°C for 2 weeks. Colonies showing the characteristic appearanceof filamentous actinomycetes were selected for isolation and the coloniesgrowing on agar plates were examined under the microscope *in situ* using the long working distance condenser and objectives. Actinomycete colonies were picked up and inoculated onto oatmeal agar slants (40 g oatmeal, 20 g agar, 1 l tap water, pH 7.2) which were incubated at 28°C for 2 – 3 weeks.

Bredholdt et al. (2008) reported to use isolation media consisted of the following: IM5 (humic acid agar, Hsu and Lockwood, 1975), with sea water), humic acid (1 g), K_2HPO_4 (0.5 g), $FeSO_4.7H_2O$ (1 mg), agar (20 g), vitamin B solution (1 ml), natural sea water (0.5 l) and distilled water (0.5 l); IM6 glycerol (0.5 g), starch (0.5 g), sodium propionate (0.5 g), KNO_3 (0.1 g), asparagine (0.1 g), casein (0.3 g), K_2HPO_4 (0.5 g), $FeSO_4.7H_2O$ (1 mg), agar (20 g), vitamin B solution (1 ml), natural sea water (0.5 L) and distilled water (0.5 l); IM7 (chitin agar, with sea water) chitin (Sigma), K_2HPO_4 (0.5 g), $FeSO_4.7H_2O$ (1 mg), agar (20 g), vitamin B solution (1 ml), natural sea water (0.7 l) and distilled water (0.3 l); IM8, malt extract (1 g), glycerol (1 g), glucose (1 g), peptone (1 g), yeast extract (1 g), agar (20 g), natural sea water (0.5 l) and distilled water (0.5 l). The pH of the isolation media was adjusted to pH 8.2. Vitamin B solution consisted of the following: thiamine-HCl (50 mg), riboflavin (50 mg), niacin (50 mg), pyridoxine-HCl (50 mg), inositol (50 mg), Ca-pantothenate (50 mg), p-aminobenzoic acid (50 mg), biotin (25 mg) and distilled water (100 ml). All isolation media were amended with filtered (0.2-μm pore size) cycloheximide (50 μg/ml), nystatin (75 μg/ml) and nalidixic acid (30 μg/ml). Seventeen media were used according to well known recipes or from known suppliers (Maldonado et al. 2009) and they were supplemented with nystatin (antifungal; 50 mg/ml) and rifampicin (5 mg/ml). Some media were also supplemented with nalidixic acid (10 mg/ml to diminish the growth of marine bacteriausing both distilled and sea water (35 g/1 l; Instant Ocean, USA).

3. Identification and Diversity of Actinobacteria

3.1. Identification Techniques

Aactinomycetes strains were taxonomic studied by employing the polyphasic approach. The isolates were subcultured onto two media, glucose-yeast-malt extract agar (ISP 2 medium; Shirling and Gottlieb 1966), and glucose-yeast extract agar (GYEA; Gordon and Mihm 1962) prepared with normal and sea water, incubated for up to 3 weeks at 28°C and then checked for purity by a light microscope (Maldonado et al., 2009). The suspensions of hyphal fragments or spores of isolates were preserved in 20% glycerol (w/v) at -20°C and -80°C for long term maintenance. Morphological grouping assigned the isolates from each sample on the basis of the properties of the colonies (i.e., colour, pigment formation, etc.) when they grew on the solid media. If aerial mycelium or spores were evident, the microorganisms were then also prepared for scanning electron microscopy (SEM) by examining gold-coated, dehydrated material, prepared from up to 28-days-old cultures.

The phenotypic characteristics of isolates compared with the type strains including morphological, cultural, physiological and biochemical characteristics such as carbon utilization, the ability to produce acid from sugars, the temperature tests and the pH tolerance test were determined by following the standard protocol of the International Streptomyces Project (ISP) (Shirling and Gottlieb 1966; Arai, 1975; Williams and Cross, 1971; Gordon et al. (1974). The chemotaxonomic characteristics including the determination of menaquinone (Collins 1977), polar lipid composition of isolate (Minnikin et al. 1984), mycolic acids (Minnikin et al. 1980), the DNA G+C mol% content (Gonzalez and Saiz-Jimenez, 2002) and DNA-DNA hybridization in microdilution-well plates, as reported by Ezaki et al. (1989) were conducted. 16S rRNA gene sequencing analyses was dertermined using the method of Kim et al. (1999) for which the chromosomal DNA, PCR amplification and direct sequencing of the PCR products of isolates were carried out. The 16S rRNA gene sequence was multiple-aligned with selected sequences obtained from the GenBank/EMBL/DDBJ databases by using CLUSTAL W version 1.81 (Thompson et al., 1997). Phylogenetic trees were reconstructed by using the neighbour-joining (Saitou and Nei, 1987), maximum-parsimony (Kluge and Farris, 1969) and maximum-likelihood (Felsenstein, 1981) methods in the program MEGA5.0 (Tamura et al., 2011). The confidence values of nodes were evaluated by using the bootstrap resampling method with 1000 replicates (Felsenstein, 1985).

3.2. Diversity of Actinobacteria

The novel actinomuycetes from marine sediments are proposed including *Verrucosispora gifhornensis* HR1-2^{T}, a new genus, isolated from a peat bog near Gifhorn, Lower Saxony, Germany (Rheims et al., 1998); *Pseudonocardia antarctica* DVS 5a1, isolated from McMurdo Dry Valleys, Antarctica (Prabahar et al., 2003); *Sciscionella marina* SCSIO 00231^{T}, gen. nov., sp. nov., isolated from a sediment in the northern South China Sea (Tian et al., 2009); *Verrucosispora sediminis* MS426^{T}, from a deep-sea sediment sample of the South China Sea, a cyclodipeptide-producing actinomycete from deep-sea sediment (Dai et al., 2010); *Streptomyces glycovorans* YIM M 10366^{T}, *Streptomyces xishensis* YIM M 10378^{T} and *Streptomyces abyssalis* YIM M 10400^{T} from marine sediments collected from the Xisha Islands in the South China Sea. (Xu et al., 2012); *Spinactinospora alkalitolerans* CXB654^{T} gen. nov. in the family *Nocardiopsaceae*, isolated from marine sediment collected at a depth of 17.5 m near the Yellow Sea Cold Water Mass, China (Chang et al., 2011); *Verrucosispora maris* AB-18-032^{T} a novel deep-sea actinomycete isolated from a marine sediment which produced abyssomicins (Goodfellow et al., 2012); *Streptomyces pharmamarensis* PM267^{T} isolated from a marine sediment, sediment sample in the Mediterranean Sea (Carro et al., 2012); *Micromonospora sediminicola* SH2-13^{T}, isolated from a marine sediment sample collected from the Andaman Sea of Thailand (Supong et al., 2013); *Verrucosispora fiedleri* MG-37, an actinomycete isolated from a fjord sediment which synthesized proximicins (Goodfellow et al., 2013); *Verrucosispora qiuiae*, isolated from mangrove swamp sediment (Xi et al., 2012); *Streptomyces chumphonensis* isolated from marine sediments in Thailand (Phongsopitanun et al., 2014) and *Micromonospora fluostatini* PWB-003^{T}, which produced fluostatins B and C antibiotics, isolated from nearshore sediment collected from Panwa Cape, Phuket Province, Thailand (Phongsopitanun et al., 2015).

The distribution of obligate marine bacteria from ocean sediments within the order *Actinomycetales* designated MAR 1 has been reported (Mincer et al., 2002). Actinobacterial populations designed as MAR 1 for *Micromonosporaceae* (*Micromonospora* and *Salinispora*); MAR2 and MAR3 and MAR4 for *Streptomycetaceae* (*Streptomyces* and *Kitasatospora*); MAR5 and MAR6 for *Thermomonosporaceae (Actinomadura*) were described by Jensen et al. (2005) based on the small subunit rRNA (*SSU rRNA) gene sequences* data analysis. Stach et al., 2003 investigated bacterial diversity in a deep-sea sediment, Atlantic ocean deep-sea sediment collected from the edge of the Saharan debris flow near the Canary Islands (27°02.39'N 18°29.02'W) at a depth of 3,814 m using a piston corer during a scientific cruise aboard the RRS Charles Darwin by constructing actinobacterium-specific 16S ribosomal DNA (rDNA) clone libraries from sediment sections taken 5 to 12, 15 to 18, and 43 to 46 cm below the sea floor at a depth of 3,814 m. Clones were placed into operational taxonomic unit (OTU) groups with >99% 16S rDNA sequence similarity; the cutoff value for an OTU was derived by comparing 16S rRNA homology with DNA-DNA reassociation values for members of the class Actinobacteria. The 5- to 12-cm sediment actinobacterium community was dominated by bacteria most closely related to *Streptomyces* species (45%), though OTUs from this sediment section were distributed throughout the phylogenetic tree. OTUs from the 15- to 18-cm sediment section were dominated by bacteria most closely related to *Rhodococcus* species (56%), with only one OTU (M16) being most closely related to a *Streptomyces* species. The 43- to 46-cm community was also dominated by *Rhodococcus* species (62%), with no *Streptomyces* species present.

Magarvey et al. (2004) reported 102 actinomycetes isolated from subtidal marine sediments collected from the Bismarck Sea and the Solomon Sea off the coast of Papua New Guinea. A combination of physiological parameters, chemotaxonomic characteristics, distinguishing 16S rRNA gene sequences, and phylogenetic analysis based on 16S rRNA genes provided strong evidence for the two new genera (represented by strains of the PNG1 clade and strain UMM518) within the family *Micromonosporaceae* including *Micromonospora* and *Verrucosispora.* Actinobacteria isolated from marine environments have been dominated by *Micromonospora*, *Rhodococcus* and *Streptomyces* species (Maldonado et al., 2005). The dominant actinomycete recovered from marine samples collected around the island of Guam belonged to the seawater-requiring marine taxon *Salinispora*, a new genus within the family *Micromonosporaceae* and two major new clades related to *Streptomyces* spp., tentatively called MAR2 and MAR3, including the new genera within the *Streptomycetaceae* and five new marine phylotypes, including two within the *Thermomonosporaceae* (Jensen et al., 2005). Phylogenetic analysis of 189 representative isolates from sediments collected in the Republic of Palau from the intertidal zone to depths of 500 m., based on 16S rRNA gene sequence data, indicated that 124 (65.6%) belonged to the class *Actinobacteria* including *Micromonosporaceae*, *Nocardiaceae*, *Nocardioidaceae* and *Streptomycetaceae*, spore-forming strains from the *Pseudonocardiaceae* and *Thermomonosporaceae* while the remaining 65 (34.4%) were members of the class *Bacilli* including *Bacillus*, *Pontibacillus*,

Paenibacillus, and *Laceyella* (Gontang et al., 2007). *Streptomyces*, *Micromonospora*, *Actinocorallia*, *Actinomadura*, *Knoellia*, *Glycomyces*, *Nocardia*, *Nocardiopsis*, *Nonomuraea*, *Pseudonocardia*, *Rhodococcus* and *Streptosporangium* genera were isolated from the shallow water sediments of the Trondheim fjord (Norway) (Bredholdt et al., 2007). *Actinomadura*, *Dietzia*, *Gordonia*, *Micromonospora*, *Nonomuraea*, *Rhodococcus*, *Saccharomonospora*, *Saccharopolyspora*, *Salinispora*, *Streptomyces*, *"Solwaraspora"* and *Verrucosispora* were isolated from marine sediments samples collected in the Gulf of California and the Gulf of Mexico (Maldonado et al., 2009). Actinobacteria from marine sediments were isolated from the Valparaíso bay, Chile, containing *Aeromicrobium*, *Agrococcus*, *Arthrobacter*, *Brachybacterium*, *Corynebacterium*, *Dietzia*, *Flaviflexus*, *Gordonia*, *Isoptericola*, *Janibacter*, *Microbacterium*, *Mycobacterium*, *Ornithinimicrobium*, *Pseudonocardia*, *Rhodococcus*, *Streptomyces*, *Tessaracoccus* and one isolate as a novel phylogenetic branch related to the *Nocardiopsaceae* family (Claverias et al., 2015). The distribution of actinobacteria in marine-sediment is summarized in Table 2.

Table 2. Distribution of actinobacteria in marine-sediments

Species	Sources/locations	References
Actinomadura sp. CNU-125	Marine sediments from Kahlisco, El Cajon, CA	Gontang et al., 2007
Amycolatopsis sp. GY109, *Frankia* sp., *Kitasatospora* sp.	Deep sea sediment (3800 m)	Stach et al., 2003
Microbacterium sp. strains CNJ-797, CnJ-743, CNJ-930	Marine sediments from Kahlisco, El Cajon, CA	Gontang et al., 2007
Micromonospora sp. strains CNB394 and CNB512	Marine sediments from Bahamas	Mincer et al., 2002
M. rhodorangea, *M. halophytica*	Marine sediment from Papu new Guinea	Magarvey et al., 2004
M. sediminicola SH2-13^T	The gulf of Thailand	Supong et al., 2013
Marinactinospora thermotolerans	Marine sediment in the northern South China Sea	Tian et al., 2009a
Nocardia sp. CnS-044	Marine sediments from Kahlisco, El Cajon, CA	Gontang et al., 2007
Pseudonocardia alaminiphila DVS 5a1^T, *P. aurantiaca*, *P. alnii*, *P. antarctica*	Deep sea sediment (3800 m) McMurdo Dry Valleys region of Antarctica	Maldonado et al., 2005b; Stach et al., 2003; Prabahar et al., 2004

Species	Sources/locations	References
Salinispora aenicola CNS-051, *S. tropica* CNS-055	Marine sediments from Bahamas	Maldonado et al., 2005a
	Marine sediments from Kahlisco, El Cajon, CA	Gontang et al., 2007
Sciscionella marina SCSIO 00231^T	Marine sediments in the northern South China Sea	Tian et al., 2009b
Spinactinospora alkalitolerans CXB654^T	Marine sediment from the Yellow Sea Cold Water Mass, China	Chang et al., 2011
Streptomyces glycovorans YIM M10366^T, *S. xishensis* YIM M 10378^T, *S. abyssalis* YIM M 10400^T	Marine sediment from the Xisha Island, China	Xu et al., 2012
S. pharmamarensis PM267^T	Marine sediment, Mediterranean Sea	Carro et al., 2012
Verrucosispora sedimidis MS426^T	Deep-sea sediment, South China Sea	Dai et al., 2010
*V. maris*AB-18-032^T	Marine sediment from the sea of Japan	Goodfellow et al., 2012

4. Secondary Metabolites and Biological Activity of Actinobacteria

4.1. Secondary Metabolites and Biological Activity of *Streptomyces*

Marine sediment actinomycete in the genus *Streptomyces* are dominantly found while the rare actinomycetes including genera *Actinoalloteichus*, *Actinomadura*, *Marinactinospora*, *Marinispora*, *Nocardiopsis*, *Saccharomonospora*, *Salinispora* and *Verrucosispora* are isolated. *Streptomyces* strains provided valuable source of new bioactive compounds and biological activity. *S. aureoverticillatus* NPS001583 from marine sediment, the strain produced aureoverticillactam, a novel 22-atom macrocyclic lactam, novel anticancer and anti-infective agents (Mitchell et al., 2004). Actinomycete (MAR4) strain CNQ-525 from ocean sediments collected at a depth of 152 m near La Jolla, California produced three new chlorinated dihydroquinones and one previously reported analogue (Soria-Mercado et al., 2005). *S. nodosus* NPS007994 from a marine sediment collected in Scripps Canyon, La Jolla, California, was found to produce lajollamycin a nitro-tetraene spiro--lactone-γ- lactam antibiotic and showed antimicrobial activity against both drug-sensitive and -resistant Gram-positive bacteria and inhibited the growth of B16-F10 tumor cells *in vitro* (Manam et al., 2005). *Streptomyces* sp. NPS008187 from a marine sediment collected in Alaska was found to produce three new pyrrolosesquiterpenes, glyciapyrroles A, B, and C, along with the known diketopiperazines cyclo(leucyl-prolyl), cyclo(isoleucyl-prolyl), and cyclo(phenylalanyl-prolyl) (Macherla et al., 2005).

Streptomyces sp. M045 produced two novel antitumor antibiotics, chinikomycins A and B including manumycin A. The compounds exhibited antitumor activity against different human cancer cell lines, but were inactive in antiviral, antimicrobial, and phytotoxicity tests (Li et al., 2005). *Streptomyces* sp. KORDI-3238 from deep-sea sediments produced a new cytotoxic compound, streptokordin, and four known compounds, nonactic acid, dilactone, trilactone, and nonactin, streptokordin. Streptokordin exhibited significant cytotoxicity against seven human cancer cell lines but showed no growth inhibition against various microorganisms including bacteria and fungi (Jeong et al., 2006). *Streptomyces* sp. QD518 produced a new staurosporinone, *N*-carboxamido-staurosporine, and a new sesquiterpene, (5*S*, 8*S*, 9*R*, 10*S*)-selina-4, 7-diene-8, 9-diol (Wu et al., 2006). A marine-derived actinomyces strain NPS554 from a marine sediment collected from Miyazaki Harbor, Japan, at a depth of 38 m yielded two trialkyl-substituted aromatic acids, lorneic acid A and lorneic acid B. Their structural differences affected inhibition activities against phosphodiesterase 5 (Iwata et al., 2006). *Streptomyces* strain CNQ-085 produced four new cytotoxic compounds, designated as daryamides A, B, and C and (2*E*, 4*E*)-7-methylocta-2,4-dienoic acid amide. The daryamides show weak to moderate cytotoxic activity against the human colon carcinoma cell line HCT-116 and very weak antifungal activities against *Candida albicans* (Asolkar et al., 2006).

Streptomyces sp. CNH990 from marine sediment produced two new cytotoxic quinones of the angucycline class, marmycins A and B. Marmycin A displayed significant cytotoxicity against several cancer cell lines, some at nanomolar concentrations; while compound B, a chloro analogue of A, was less potent. For marmycin A, tumor cell cytotoxicity appeared to coincide with induction of modest apoptosis and arrest in the G1 phase of the cell cycle (Martin et al., 2007). *Streptomyces* sp. KORDI-3973 isolated from the deep sea sediment produced streptopyrrolidine, abenzyl pyrrolidine derivative that exhibited significant anti-angiogenesis activity (Shin et al., 2008).

Streptomyces sp. CNQ-418 furnished the marinopyrroles A and B that possess potent antibiotic activities against methicillin resistant *Staphylococcus aureus* (Hughes et al., 2008). *Streptomyces.* sp. CNQ-617 produced two novel spiroaminals, marineosins A and B, containing two pyrrole functionalities, showed significant inhibition of human colon carcinoma (HCT-116) in an in vitro assay (IC_{50}) 0.5 μM for marineosin A) and selective activities in diverse cancer cell types (Boonlarppradab et al., 2008). *Streptomyces* sp. 17944 produced three new tirandamycins (TAMs), TAM E, F, and G, along with TAM A and B, that 5 selectively inhibits the *Brugia malayi* AsnRS and efficiently kills the adult *B. malayi* parasite, representing a new lead scaffold to discover and develop antifilarial drugs.

Streptomyces sp. 307-9 produced the novel dienoyl tetramic acids tirandamycin C and tirandamycin D with activity against vancomycin-resistant *Enterococcus faecalis* and also the previous compounds tirandamycins A and B (Carlson et al., 2009; Yu et al., 2011). *Streptomyces albidoflavus* NTK 227 from Atlantic Ocean sediment produced albidopyrone, a new -pyrone-containing secondary metabolite and found to have a moderate inhibitory activity against protein-tyrosin phosphatase B (Hohmann et al., 2009a). *Streptomyces* sp. NTK 937 produced caboxamycin, a new benzoxazole antibiotic which inhibited against Gram-positive bacteria, selected human tumor cell lines and the enzyme phosphodiesterase (Hohmann et al., 2009b). *Streptomyces* sp. Sp080513GE-23 isolated from a sponge-derived actinomycete, produced tetrapeptides possessing a unique skeleton, JBIR-34 and JBIR-35 (Motohashi, et al., 2010). Actinomycete strain CNQ-509 produced five new farnesyl-*R*-nitropyrroles, nitropyrrolins A E, several of the nitropyrrolins, nitropyrrolin D in particular, are cytotoxic toward HCT-116 human colon carcinoma cells, but show weak to little antibacterial activity against methicillin-resistant *Staphylococcus aureus* (MRSA) (Kwon et al., 2010). *Streptomyces tumescens* YM23-260 produced two peptides, tumescenamides A and B. Tumescenamide A induced reporter gene expression under the control of the insulin-degrading enzyme promoter (Motohashi et al., 2010).

Streptomyces sp. CNQ-027 produced three highly modified peptides, actinoramides A C (Nam et al., 2011). *Streptomyces antibioticus* H74-18 produced antimycins A_{19} and A_{20}, two new antimycins. All the antimycins showed potential antifungal activities against *Candida albicans* with MIC of about 5 – 10 μg/ml (Xu et al., 2011). *Streptomyces* sp. CNS-575 produced fijimycins A – C, three antibacterial etamycin-class depsipeptides. Fijimycins

A – C, and etamycin A, were shown to possess significant in vitro antibacterial activity against three methicillin-resistant *Staphylococcus aureus* (MRSA) strains with MIC 100 values between 4 and 16 μg/ml (Sun et al., 2011). *Streptomyces* sp. SCSIO 01127, isolated from the South China Sea sediment, produced two new spirotetronate antibiotics lobophorins E and F, along with two known analogs lobophorins A and B. The new compound lobophorin F showed antibacterial activities against *Staphylococcus aureus* ATCC 29213 and *Enterococcus faecalis* ATCC 29212 with MIC values of 8 μg/ml for both the strains, better than that of lobophorin B. Lobophorin F also displayed better cytotoxic activities than lobophorin B, with IC_{50} of 6.82, 2.93 and 3.16 μM against SF-268, MCF-7 and NCI-H460, respectively (Niu et al., 2011). *Streptomyces* sp. NPS853 from marine sediments, it produced new anthramycin-type analogues, designated usabamycin A – C. The usabamycins show weak inhibition of HeLa cell growth and selective inhibition of serotonin (5-hydroxytrypamine) 5-HT2B uptake (Sato et al., 2011).

Streptomyces spinoverrucosus SNB-032 produced four new anthraquinone analogues including galvaquinones A – C and an isolation artifact, 5,8-dihydroxy-2,2,4-trimethyl-6-(3-methylbutyl)- anthra[9,1-de][1,3]oxazin-7(*2H*)-one. Galvaquinone B was found to show epigenetic modulatory activity at 1.0 μM and exhibited moderate cytotoxicity against non-small-cell lung cancer (NSCLC) cell lines Calu-3 and H2887 (Hu et al., 2012). *Streptomyces lusitanus* SCSIO LR32, an actinomycete of deep sea origin produced Five new *C*-glycoside angucyclines, named grincamycins B – F, and a known angucycline antibiotic, grincamycin, grincamycin (Huang et al., 2012). A deep-sea-derived *Streptomyces* sp. SCSIO 03032 produced a new bisindole alkaloids spiroindimicins A D. Spiroindimicins B D with a [5, 5] spiro-ring exhibited moderate cytotoxicities against several cancer cell lines (Zhang et al., 2012). *Streptomyces* sp. WBF-16 produced two new anthraquinone glycosides strepnoneside A and strepnoneside B, together with chromomycin A3. Chromomycin A3 exhibited cytotoxic activities against HCT 116 cell lines ($IC_{50} = 300 \pm 11$ *p*M) (Lu et al., 2012). *Streptomyces* sp. FMA produced streptocarbazoles A and B, two novel indolocarbazoles. Streptocarbazoles A was cytotoxic on HL-60 and A-549 cell lines and could arrest the cell cycle of Hela cells at the G2/M phase (Fu et al., 2012).

Streptomyces sp. CNQ-027 isolated from marine sediment sample collected at a depth of 50 m of the coast of Diego, CA produced a new meroterpenoid, actinoranone that exhibited significantly cytotoxic to HCT-116 human colon cancer cells with an LD_{50} 2.0 μg/ml, anti-cancer and tumor cell lines (Nam et al., 2013). *Streptomyces* sp. TP-A0867 from marine sediment sample collected at a depth 38 mnear Miyazaki Harbor, Japan, produced akaeolide, a novel polycyclic polyketide, anti-bacteria and cytotoxicity (Igarashi et al., 2013). *Streptomyces* sp. CNH-287 produced chlorizidine, a cytotoxic 5*H*pyrrolo [2,1*a*]isoindol-5-one containing alkaloid, cytotoxicity against HCT-116 human colon cancer cells (Alvarez-Mico et al., 2013). *Streptomyces* sp. 7-145, had the potential to produce glycosidic antibiotic, an elaiophylin derivatives showed good antibacterial activity against methicillin-resistant *Staphylococcus aureus* and vancomycin-resistant enterococci pathogens (Wu et al., 2013). *Streptomyces* sp. CNT-372 from Fiji produced farnesides A and B, sesquiterpenoid nucleoside ethers. Farneside A exhibited modest antimalarial activity against the parasite *Plasmodium falciparum* (Iland et al., 2013). *Streptomyces niveus* SCSIO 3406 from a South China Sea sediment sample obtained from a depth of 3536 m produced four new sesquiterpenoid naphthoquinones, marfuraquinocins A – D, and two new geranylated phenazines, phenaziterpenes A and B. Marfuraquinocins exhibited antibacterial activities against *Staphylococcus aureus* ATCC 29213 (Song, et al., 2013). *Streptomyces* sp. RJA2928 obtained from a tropical marine sediment, produced nahuoic acid A, a selective SAM-competitive inhibitor of the histone methyltransferase SETD8 (Williams et al., 2013). In addition, nahuoic acids B E were produced by the marine *Streptomyces* sp. SCSGAA 0027 (Nong et al., 2016), and were found in the strain RJA2928 (Williams et. al., 2016). Nahuoic acid B E showed weak antibiofilm activity against Shewanella onedensis MR-1 biofilm. *Streptomyces* sp. SNJ013 isolated from deep-sea sediment collected off Jeju Island, Korea, produced sungsanpin, a new 15-amino-acid peptide. Sungsanpin displayed inhibitory activity in a cell invasion assay with the human lung cancer cell line A549 (Um et al., 2013). *Streptomyces* sp. JAMM992 produced surugamides A – E, cyclic octapeptides with four *D*amino acid residues (Takada et al., 2013). The metabolites and biological activity of *Streptomycetes* from marine sediments that published on 2010-2016, are summarized in Table 3.

Table 3. Secondary metabolites and biological activity of streptomycetes

Compound	Structural class	Biological activity	Producer	Source/location	References
Actinoramides A-C	Peptide	Not detected	*Streptomyces sp.* CNQ-027	Marine sediment sample collected at a depth of 50 moff the coast of San Diego, CA	Nam et al., 2011
Actinoranone	Meroterpenoid	Anti-cancer and tumor cell lins	*Streptomyces* sp. CNQ-027	Marine sediment sample collected at a depth of 50 m of the coast of Diego, CA	Nam et al., 2013
Akaeolide	Carbocyclic polyketide	Anti-bacteria and cytotoxicity	*Streptomyces* sp. NPS554	Marine sediment sample collected at a depth 38 m near Miyazaki Harbor, Japan	Igarashi et al., 2013
Antimycins A19-A20	3-formami-dosalicylic acid-amide link dilactone ring	Anti-fungal (*C. albicans*) activity	*Streptomyces antibioticus* H74-18	Marine sediment collected at a mangrove zone in the South China Sea, Guangdong province, China.	Xu et al., 2011
Chlorizidine	Chlorinated nitrogen-containing carbon skeleton	Anti-cancer and cytotoxicity	*Streptomyces* sp. CNH-287	Marine sediment sample collected at the intertidal zone near San Clemente, CA	Alvarez-Mico et al., 2013
Elaiophylin derivatives	Lactone macrolide glycoside	Anti-bacteria	*Streptomyces* sp. 7-145	Marine sediment sample collected at a depth of 40 m. from Heishijiao Bay, Dalian, China	Wu et al., 2013
Farnesides	Sesquiterpenoid Nucleoside ether	Anti-malaria, anti-bacteria and anti-influenza type A virus	*Streptomyces* sp. CNT-372	Marine sediment sample at a depth of 5 m, Nacula Island in the Yasawa Island chain, Fiji	Iland et al., 2013
Fijimycins A- C	Depsipeptide	Anti-bacterial (MRSA) activity	*Streptomyces* sp. CNS-575	Marine sediment samples collected at a depth of 0.5 m from the Nasese shoreline,	Sun et al., 2011

Compound	Structural class	Biological activity	Producer	Source/location	References
				Viti Levu, Fiji	
Galvaquinones A- C	Anthraquinone	Anti-cancer activity	*Streptomyce spinoverrucosu* SNB-032	Marine sediment sample collected from Trinity Bay, Galveston, Texas	Hu et al., 2012
Glucopiericidin C	Piericidin glucoside	Anti-cancer and cytotoxicity	*Streptomyces* sp. stB8112	Marine sediment collected from the Laguna deTerminos in the Gulf of Mexico	Shaaban et al., 2011
Grincamycins B-F	Anthraquinone glycoside	Anti-cancer activity	*Streptomyces lusitanu* SCSIO LR32	Marine sediment sample at a depth of 3370 m. collected in the South China Sea	Huang et al., 2012
JBIR-34 and 35	Tetrapeptide	Anti-oxidant (anti-DPPH redical) and anti-cancer cell lines	*Streptomyces* sp. Sp080513GE-23	Marine sponge, Haliclona sp., collected from Tateyama city, Chiba Prefecture, Japan.	Motohashi, et al., 2010
Lobophorins E- F	Spirotetronate glycoside	Anti-bacterial and anti-cancer activities	*Streptomyces* sp. SCSIO 01127	Marine sediment sample at the depth of 1350 m in the South China Sea	Niu et al., 2011
Marfuraquinocins A-D	Sesquiterpenoid naphthoquinone	Anti-bacterial and anti-cancer cell lines	*Streptomyces niveus* SCSIO 3406	Marine sediment sample at a depth of 3536 m. collected in the South China Sea	Song, et al., 2013
Nahuoic acids	Polyketide	Anti- histone methyltransferase SETD8	*Streptomyces* sp. RJA2928 and SCSGAA 0027	Marine sediment collected near the passage Padana Nahua in Papua New Guinea	Williams et al., 2013; Nong et al., 2016; Williams et al., 2016;
Nitropyrrolins A-E	Farnesyl-α-nitropyrrole	Anti-bacteria and anti-cancer cell lines	*Streptomyces* sp. CNQ-509	Marine sediment collected at a depth of 44 m off shore of La Jolla, CA.	Kwon et al., 2010

Table 3. (Continued)

Compound	Structural class	Biological activity	Producer	Source/location	References
Spiroindimicins A-D	Bisindol alkaloid	Anti-cancer activity	*Streptomyces* sp. SCSIO 03032	Marine sediment sample of the India Ocean at the depth of 3412 m	Zhang et al., 2012
Strepnonesides	Anthraquinone glycoside	Anti-cancer activity	*Streptomyces* sp. WBF-16	Marine sediments sample collected from Wei-hai Sea, China	Lu et al., 2012
Streptocarbazoles A-B	Indolocarbazole	Anti- Hela cells at the G2/M phase	*Streptomyces* sp. FMA	Marine sediments collected from the protected areas of mangrove at Sanya, China	Fu et al., 2012
Sungsanpin	Peptide	Anti-cancer cell lines	*Streptomyces* sp. SNJ013	Marine sediments collected at depths of 108-138 m. from the sea floor off the coast of Sungsanpo on Jeju Island, Republic of Korea	Um et al., 2013
Surugamides A-E	Cyclic octapeptide	Anti- bovine cathepsin B and anti-cysteine protease	*Streptomyces* sp. JAMM992	Marine sediment collected from Kinko Bay, Japan	Takada et al., 2013
Tirandamycins	Dienoyl tetramic acid	Anti-*Brugia malayi* asparagine tRNA synthetase and anti-vancomycin-resistant *Enterococcus faecalis*	*Streptomyces* sp. 307-9 and 17944	Marine sediments collected at a depth of 15 m during a scuba expedition in Salt Cay, U.S.	Carlson et al., 2009; Yu et al., 2011
Tumescenamide A-B	Peptide	Control of the insulin-degrading enzyme promoter	*Streptomyce tumescen* M23-260	Marine sediments of Big drop-off of the Republic of Palau	Motohashi et al., 2010
Usabamycins A-C	Pyrrolobenzo-diazepine	Anti- f HeLa cell and anti-serotonin (5-hydroxyt -rypamine) 5-HT2B	*Streptomyces* sp. NPS853	Marine sediment sample collected at a depth of 20 m. Usa Bay, Kochi Prefecture, Japan	Sato et al., 2011

4.2. Secondary Metabolites and Biological Activity of Rare Actinomycetes

Rare actinomycetes are promising sources in search for new drugs, and their potential for producing biologically active molecules, were studied (Bredholdt et al., 2007). A wide variety of secondary metabolites from marine sediment actinomycetes were used in clinical therapy, and some compounds were originally guided for chemical synthesis such as salinosporamide A, isomeric abyssomicin C, and marinomycin A (Endo and Danishefsky et al., 2005; Nicolaou and Harrison, 2007; Nishimaru et al., 2014). For marine natural products, salinosporamide A is a chlorinated polyketide metabolite that belongs to the class of -lactones. The compound is a potent proteasome inhibitor being used as the anticancer agent. It was produced by obligated marine actinomycete, *Salinispora tropica* which was originally isolated from marine sediments (Feling et al., 2003). This compound is currently being evaluated in phase I clinical trials as monotherapy and in combination with dexamethasome in patients with relapsed/refractory MM. Furthermore, the combination therapy of salinosporamide A studies bearing minimally cytotoxic synergistic effects in MM, leukemia, and lymphoma cell lines, this combination significantly decreased viability in tumor cells from five relapsed MM patients and reduced tumor growth in human MM xenograft mouse model without any noticeable toxicity (Chauhan et al., 2008; Gulder and Moore, 2010).

Atrop-abyssomicin C is the polycyclic polyketide antibiotic that is atropisomer of abyssomicin C. This abyssomicin isomer is produced by the strain in genus *Verrucosispora* (Bister et al., 2004; Riedlinger et al., 2004; Keller et al., 2007). For the abyssomicin C initially discovered with activity against -aminobenzoate (ABA) in Gram-positive bacteria *Bacillus subtilis* and *Staphylococcus aureus* (Bister et al., 2004; Riedlinger et al., 2004), in addition, it showed *in vivo* activity against *Mycobacterium tuberculosis* (Freundlich et al., 2010). Marinomycin A is potent antitumor agent with substantial activities against selected human tumor cells and drug-resistant pathogenic bacteria. This compound derived from marine actinomycete *Marinispora* that isolated from the marine sediments (Kwon et al., 2006). Marinomycin A has highly enhanced *in vitro* activity againstsix of eight melanoma cell lines, with SK-MEL-5 showing the highestsensitivity (concentration lethal to 50% of animals tested that showed the LC_{50} at 5.0 *n*M. Marinomycin A also inhibits the growth of human pathogenic bacteria with a minimum inhibitory concentration (MIC) valueof 0.1 μM against methicillin-resistant *Staphylococcus aureus* (MRSA) and vancomycin-resistant *Enterococcus faceium* (VREF). Unfortunately, these polyenes are highly photoreactive and undergo polyene isomerization under room light. This reactivity and general long-term instability detracts from their potential for clinical development. The chemical structure of bioactive substances from marine sediment-derive actinomycetes was shown in Figure 1.

Recently, the diversity of rare actinomycetes from sediments and the secondary metabolites with diverse biological activities have been investigated (Bredholdt et al., 2007).

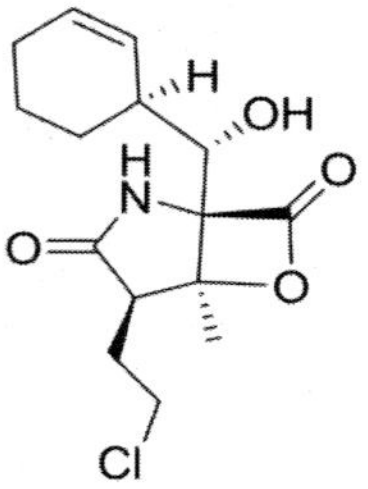

Salinosporamide A

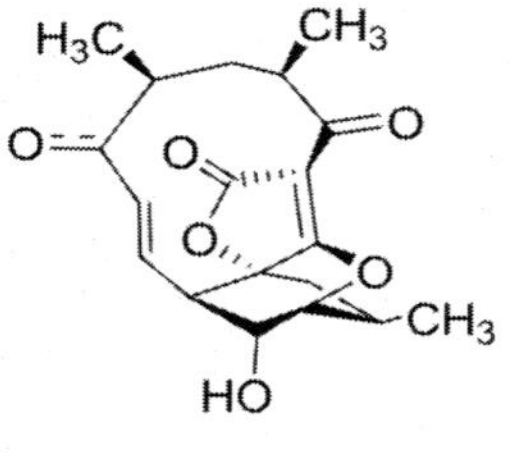

Atrop-abyssomicin C

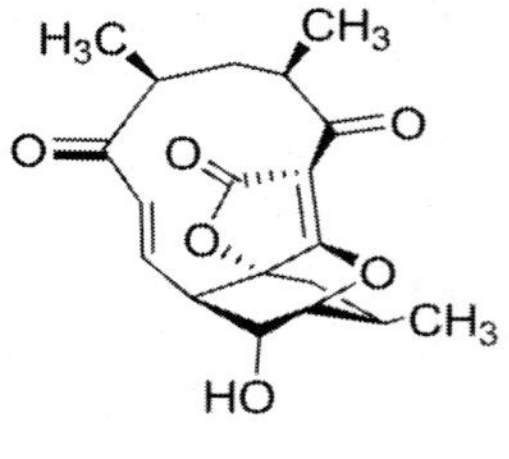

Abyssomicin C

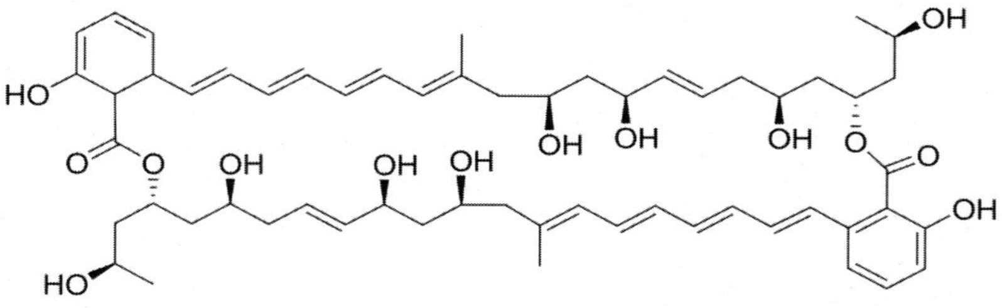

Marinomycin A

Figure 1. Chemical structure of bioactive metabolites from marine actinomycetes.

Marinactinospora thermotolerans SCSIO 00652 from the deep South China Sea was found to produce a new sequential tristhiazole-thiazoline-containing cyclic peptide, marthiapeptide A that exhibited antibacterial activity against a panel of Gram-positive bacteria, with MIC values ranging from 2.0 to 8.0 μg/mL, and displayed strong cytotoxic activity against a panel of human cancer cell lines with IC50 values ranging from 0.38 to 0.52 µM (Zhou et al., 2012). This strain produced four new β-carboline alkaloids, designated marinacarbolines AD, two new indolactam alkaloids, 13-*N*-demethyl-methylpendolmycin and methylpendolmycin-14-*O*-α-glucoside, and the three known compounds 1-acetyl-β-carboline, methylpendolmycin, and pendolmycin. The new alkaloid compounds were inactive against a panel of eight tumor cell lines (IC_{50} > 50 μM) but exhibited antiplasmodial activities against Plasmodium falciparum lines 3D7 and Dd2, with IC_{50} values ranging from 1.92 to 36.03 μM (Huang et al., 2011).

Micromonospora rosaria SCSIO N160 from a South China Sea sediment was found to produce three new fluostatins, I – K together with six known compounds, fluostatins C – F, rabelomycin and phenanthroviridone. Rabelomycin and phenanthroviridone exhibited good antimicrobial activities against *Staphylococcus aureus* ATCC 29213 with MIC values of 1.0 and 0.25 μg/ml, respectively. Phenanthroviridone also exhibited significant in vitro cytotoxic activities toward SF-268 (IC_{50} 0.09 μM) and MCF-7 (IC_{50} 0.17 μM) (Zhang et al., 2012). The marine *Actinomadura* sp. 007 produced the compound ZHD-0501, a novel naturally occurring staurosporine analog showing anticancer activity *in vitro* (Han et al., 2005).

Actinoalloteichus cyanogriseus WH1-2216-6, a mutant could produce three new acyclic bipyridine glycosides, cyanogrisides E – G, and a known cyanogriside H. Cyanogrisides F and G showed cytotoxicities against HCT116 and HL-60 cells, in addition all cyanogriside derivatives showed cytotoxic on K562 cells (Fu et al., 2011). *Actinoalloteichus* sp. NPS702 from marine sediment collected from Usa Bay, Kochi Prefecture, Japan, the strain produced nine new 26-membered macrolides of the oligomycin subfamily, neomaclafungins A – I. These compounds exhibited significant antifungal activity *in vitro* against *Trichophyton mentagrophytes* ATCC 9533 with MIC values between 1 and 3 μg/ml (Sato et al., 2012). *Marinispora* sp. NPS12745 from a marine sediment collected off the coast of San Diego, California. Strain NPS12745 produced a series of chlorinated bisindole pyrroles, lynamicins A E, which showed broad-spectrum biological activity against both Gram-positive and Gram-negative bacteria. Significantly, these compounds were active against drug-resistant pathogenic bacteria such as methicillin-resistant

Staphylococcus aureus and vancomycin-resistant *Enterococcus faecium* (McArthur et al., 2008). Marine-derived *Marinispora* sp. NPS008920 from a sediment collected in Cocos Lagoon, Guam, produced a series of novel 2-alkylidene- 5-alkyl-4-oxazolidinones, lipoxazolidinone A, B, and C. Lipoxazolidinones A – C showed broad spectrum antimicrobial activity similar to that of the commercial antibiotic linezolid (Zyvox), a 2-oxazolidinone. Hydrolysis of the amide bond of the 4-oxazolidinone ring of A resulted in loss of antibacterial activity. The 2-alkylidene-4-oxazolidinone represents a new antibiotic pharmacophore and is unprecedented in nature (Macherla et al., 2007).

Saccharomonospora sp. CNQ490 from marine sediments collected at the mouth of the La Jolla Submarine Canyon, produced the unprecedented alkaloid lodopyridone. This compound is cytotoxic to HCT-116 human colon cancer cells with IC_{50}, 3.6 μM (Maloney et al., 2009). For the obligae marine actinomycete, *Salinispora tropica* CNB-392 produced the potent proteasome inhibitor salinosporamide A and their seven related γ-lactams. The most important of these compounds were a decarboxylated pyrrole analogue, salinosporamide B. The new *structure-activity relationships* (SAR) data for eight compounds, derived from extensive testing against the human colon carcinoma HCT-116 and the 60-cell-line panel at the NCI, indicate that the chloroethyl moiety plays a major role in the enhanced activity of salinosporamide A (Felling et al., 2003; Williams et al., 2005). *Salinispora* sp. CNS103 produced cyanosporasides A and B, chloro- and cyano-cyclopenta[a]indene glycosides (Oh et al., 2006). *Salinispora arenicola* CNR-005 has led to the isolation of two unusual bicyclic polyketides, saliniketals A and B. Saliniketals A and B were found to inhibit ornithine decarboxylase induction, an important target for the chemoprevention of cancer with IC_{50} values of 1.95 0.37 and 7.83 1.2 μg/ml, respectively (Williams et al., 2007). *Salinispora arenicola* strain produced three new macrolide polyketides designated arenicolides A C (Williams et al., 2007). *Salinispora pacifica* strain led to the discovery of four new polyketides, salinipyrones A B, and pacificanones A and B (Oh et al., 2008). *Verrucosispora* sp. MG-37 produced proximicin A, B and C, novel aminofuran antibiotic that showed anticancer compounds. Proximicins showed a weak antibacterial activity but a strong cytostatic effect to various human tumor cell lines (Fiedler et al., 2008). *Verrucosispora* sp. AB-18-032 could be produced abyssomicins G and H and atrop-Abyssomicin C (Keller et al., 2007).

Table 4. Secondary metabolites and biological activity of rare actinomycetes

Compound	Structural class	Biological activity	Producer	Source/location	References
Cyanogrisides A-D	Cyclic bipyridine glycoside	Anti-cancer activity	*Actinoalloteichus cyanogriseus* WH1-2216-6	Marine sediments collected from the seashore of Weihai, China	Fu et al., 2011
Fijiolides	Chloroaromatic glycoside	Induce quinone reductase-1 (QR1) and anti-cancer activity	*Nocardiopsis* sp. CNS-653	Marine sediment sample collected from the Beqa Lagoon, Fiji	Nam et al., 2010
Fluostatins I-K	Polyketide	Anti-bacterial and anti-cancer activities	*Micromonospora rosaria*	Marine sediment sample at a depth of 30 m. collected from a South China Sea	Zhang et al., 2012
Marthiapeptide A	Cyclic peptide	Anti-bacterial and anti-cancer activities	*Marinactinospora thermotolerans* SCSIO 00652	Marine sediment sample collected in the south China Sea	Zhou et al., 2012
Marinacarbolines A-D	β-Carboline alkaloid	Anti-malarial and anti-tumor activities	*Marinactinospora thermotolerans* sSCSIO 00652	Marine sediment sample collected at a depth of 3865 m. in the northern South China Sea	Huang et al., 2011
Marinactinones A-C	γ-Pyrones	Anti-cancer activity	*Marinactinospora thermotolerans* SCSIO 00606	Marine sea sediment sample collected in the northern South China Sea	Wang et al., 2011
Neomaclafungins A-I	Lactonemacrolide	Anti-fungal activity	*Actinoalloteichus* sp. NPS702	Marine sediment collected from Usa Bay, Kochi Prefecture, Japan	Sato et al., 2012

Table 4. (Continued)

Compound	Structural class	Biological activity	Producer	Source/location	References
Nocardiamides A-B	Cyclohexapeptide	Anti-acetylcholinesterase (AChE) and anti-bacterial activities	*Nocardiopsis* sp. CNQ115	Marine sediment sample collected at a depth between 18 and 30 m near the La Jolla Canyon, San Diego	Wu et al., 2013
Nocardiopsins C-D	Macrolactam polyketide	Anti-bacterial, anti-fumgal and anti-tumor activities	*Nocardiopsis* sp. CMB-M0232	Marine sediment collected off the coast of South Molle Island, Queensland, Australia	Raju et al., 2013
Nocapyrones E-G	α-Pyrone	Anti-bacterial activity	*Nocardiopsis dassonvillei* HR10-5	Marine sediments collected from the estuary of Yellow River, Dongying, People's Republic of China	Fu et al., 2011

Nocardiopsis dassonvillei HR10-5 produced three new α-pyrones, nocapyrones E – G, and three new diketopiperazine derivatives, nocazines A – C, together with a new oxazoline compound, nocazoline A. Nocapyrones E – G showed modest antimicrobial activity against *Bacillus subtilis* with MIC values of 26, 14, and 12 μM, respectively (Fu et al., 2011). *Nocardiopsis lucentensis* CNR-712 produced four new 3-methyl-4-ethylideneproline-containing peptides, lucentamycins A D. Lucentamycins A and B showed significant *in vitro* cytotoxicity against HCT- 116 human colon carcinoma (Cho et al., 2007). *Nocardiopsis* strain CNS-653 produced fijiolide A, a potent inhibitor of TNF-R-induced NFκB activation, along with fijiolide B. Fijiolide A is viewed as a promising lead for more advanced anticancer testing (Nam et al., 2010). *Nocardiopsis* sp. CNQ115 from marine sediments collected off the coast of southern California, produced two new 4-aminoimidazole alkaloids, nocarimidazoles A and B (Leutou et al., 2015). *Nocardiopsis* sp. CMB-M0232from marine sediment collected off the coast of South Molle Island, Queensland, Australia, yielded two new examples of rare prolinyl-macrolactam polyketides, nocardiopsins C and D, a new highly substituted a-pyrone polyketide, nocardiopyrone A, and the previously reported macrolide polyketides nocardiopsins A and B. PCR amplification of CMB-M0232 genomic DNA revealed the presence of type I and type II polyketide synthase and nonribosomal peptide synthase domains (Raju et al., 2013). The metabolites and biological activity of rare actinomycetes from marine sediments that published on 2010-2016, are summarized in Table 4.

In conclusion, actinobacteria can be recovered from marine sediments like in the terrestrial soils and play important roles in the decomposition of recalcitrant organic matter in the sea floor. They become recognized as a source of novel antibiotics and anti-cancer agents. A rich biodiversity of culturable actinobacteria recovered from marine sediments belonged to genera *Streptomyces*, *Micromonospora*, *Actinocorallia*, *Actinomadura*, *Knoellia*, *Glycomyces*, *Nocardia*, *Nocardiopsis*, *Nonomuraea*, *Pseudonocardia*, *Rhodococcus*, *Streptosporangium* and *Salinispora*. *Streptomyces* strains produced diverse core structure of metabolites such as peptides, meroterpenoid, carbocyclic polyketide, chlorinated nitrogen-containing carbon skeleton, lactonemacrolide, sesquiterpenoid nucleoside ether, depsipeptide, piericidin glucoside, anthraquinone, spirotetronate glycoside, sesquiterpenoid-naphthoquinone, farnesyl-α-nitropyrrole, bisindole alkaloid, indolocarbazole, dienoyl tetramic acid, pyrrolobenzodiazepine while the rare actinomycetes, *Actinoalloteichus*, *Marinactinospora*, *Micromonospora*, *Actinoalloteichus*, *Saccharomonospora*, *Salinispora*, and *Nocardiopsis* strains produced cyclic bipyridine glycoside, chloroaromatic glycoside, cyclic peptide, *β*-carboline alkaloid, *α*- and *γ*-pyrones, macrolactone, and lactam polyketide bioactive compounds. To investigate the marine-derived actinomycetes in the world of diverse ecological habitats as the resource for biotechnology, one must understand the extent to which they are capable of growth in the ocean, the degree to which they display specific marine adaptations and the extent to which these adaptations have affected secondary metabolite production.

REFERENCES

Ahmed, L., Jensen, P. R., Freel, K. C., Brown, R., Jones, A. L., Kim, B. Y. & Goodfellow, M. (2013). *Salinispora pacifica* sp. nov., an actinomycete from marine sediments. *Antonie van Leeuwenhoek.*, *103*, 1069-1078.

Alvarez-Mico, X., Jensen, P. R., Fenical, W. & Hughes, C. C. (2013). Chlorizidine, a cytotoxic 5*H*pyrrolo [2, 1*a*] isoindol-5-one containing alkaloid from a marine *Streptomyces* sp. *Org. Lett.*, *15*, 988-991.

Arai, T. (1975). *Culture Media for Actinomycetes*. Tokyo: The Society for Actinomycetes, Japan.

Asolkar, R. N., Jensen, P. R., Kauffman, C. A. & Fenical, W. (2006). Daryamides A-C, weakly cytotoxic polyketides from a marine-derived actinomycete of the genus *Streptomyces* strain CNQ-085. *J. Nat. Prod.*, *69*, 1756-1759.

Bérdy, J. (2005). Bioactive microbial metabolites: a personal view. *J. Antibiot.*, *58*, 1-26.

Bister, B., Bischoff, D., Strbele, M., Riedlinger, J., Reicke, A., Wolter, F., Bull, A. T., Zhner, H., Fiedler, H. P. & Sussmuth, R. D. (2004). Abyssomicin C-A polycyclic antibiotic from a marine *Verrucosispora* strain as an inhibitor of the *p*-aminobenzoic acid/tetrahydrofolate biosynthesis pathway. *Angew. Chem. Int. Ed.*, *43*, 2574-2576.

Blunt, J. W., Copp, B. R., Munro, M. H. G., Northcote, P. T. & Prinsep, M. R. (2007). Marine natural products. *Nat. Prod. Rep.*, *24*, 31-86.

Boonlarppradab, C., Kauffman, C. A., Jensen, P. R. & Fenical, W. (2008). Marineosins A and B, cytotoxic spiroaminals from a marine-derived actinomycete. *Org. Lett.*, *10*, 5505-5508.

Bredholt, H., Fjærvik, F., Johnsen, G. & Zotchev, S. B. (2008). Actinomycetes from sediments in the Trondheim Fjord, Norway: Diversity and biological activity. *Mar. Drugs.*, *6*, 12-24.

Bredholdt, H., Galatenko, O. A., Engelhardt, K., Fjærvik, E., Terekhova, L. P. & Zotchev, S. B. (2007). Rare actinomycete bacteria from the shallow water sediments of the Trondheim fjord, Norway: isolation, diversity and biological activity. *Environ. Microbiol.*, *9*, 27562764.

Bulina, T. I., Alferova, I. V. & Terekhova, L. P. (1997). A novel approach to isolation of actinomycetes involving irradiation of soil samples with microwaves. *Microbiology.*, *66*, 231-234.

Bulina, T. I., Terekhova, L. P. & Tyurin, M. V. (1998). Use of electric pulses for selective isolation of actinomycetes from soil. *Microbiology.*, *67*, 459-462.

Carlson, J. C., Li, S., Burr, D. A. & Sherman, D. H. (2009). Isolation and characterization of tirandamycins from a marine-derived *Streptomyces* sp. *J. Nat. Prod.*, *72*, 2076-2079.

Carro, L., Zuniga, P., Calle, F. & Trujillo, M. E. (2012). *Streptomyces pharmamarensis* sp. nov. isolated from a marine sediment. *I Int. J. Syst. Evol. Microbiol.*, *62*, 1165-1170.

Chang, X., Liu, W. & Zhang, X. H. (2011). *Spinactinospora alkalitolerans* gen. nov., sp. nov., an actinomycete isolated from marine sediment. *Int. J. Syst. Evol. Microbiol.*, *61*, 2805-2810.

Cho, J. Y., Williams, P. G., Kwon, H. C., Jensen, P. R. & Fenical, W. (2007). Lucentamycins A-D, cytotoxic peptides from the marine-derived actinomycete *Nocardiopsis lucentensis*. *J. Nat. Prod.*, *70*, 1321-1328.

Chauhan, D., Singh, A., Brahmandam, M., Podar, K., Hideshima, T., Richardson, P., Munshi, N., Palladino, M. A. & Anderson, K. C. (2008). Combination of proteasome inhibitors bortezomib and NPI-0052 trigger *in vivo* synergistic cytotoxicity in multiple myeloma. *Blood.*, *111*, 1654-1664.

Claverías, F. P., Undabarrena, A., González, M., Seeger, M. & Cámara, B. (2015). Culturable diversity and antimicrobial activity of Actinobacteria from marine sediments in Valparaíso bay, Chile. *Front. Microbiol.*, *6*, 1-11.

Collins, M. D., Pirouz, T., Goodfellow, M. & Minnikin, D. E. (1977). Distribution of menaquinones in actinomycetes and corynebacteria. *J. Gen. Microbiol.*, *100*, 221-230.

Dai, H. Q., Wang, J., Xin, Y. H., Pei, G., Tang, S. K., Ren, B., Ward, A., Ruan, J. S., Li, W. J. & Zhang, L. X. (2010). *Verrucosispora sediminis* sp. nov., a cyclodipeptide-producing actinomycete from deep-sea sediment. *Int. J. Syst. Evol. Microbiol.*, *60*, 1807-1812.

Ezaki, T., Hashimoto, Y. & Yabuuchi, E. (1989). Fluorometric deoxyribonucleic acid-deoxyribonucleic acid hybridization in microdilution wells as an alternative to membrane filter hybridization in which radioisotopes are used to determine genetic relatedness among bacterial strains. *Int. J. Syst. Bacteriol.*, *39*, 224-229.

Endo, A. & Danishefsky, S. J. (2005). Total synthesis of salinosporamide A. *J. Am. Chem. Soc.*, *127*, 8298-8299.

Felsenstein, J. (1981). Evolutionary trees from DNA sequences: a maximum likelihood approach. *J. Mol. Evol.*, *17*, 368-376.

Felsenstein, J. (1985). Confidence limits on phylogenies: an approach using the bootstrap. *Evolution.*, *39*, 783-791.

Fiedler, H. P., Bruntner, C., Bull, A. T., Ward, A. C., Goodfellow, M., Potterat, O., Pudar, C. & Mihm, G. (2005). Marine actinomycetes as a source of novel secondary metabolites. *Antonie Van Leeuwenhoek.*, *87*, 37-42.

Feling, R. H., Buchanan, G. O., Mincer, T. J., Kauffman, C. A., Jensen, P. R. & Fenical, W. (2003). Salinosporamide A: A highly cytotoxic proteasome inhibitor from a novel microbial source, a marine bacterium of the new genus *Salinospora*. *Angew. Chem.*, *42*, 355-357.

Fiedler, H. P., Bruntner, C., Riedlinger, J., Bull, A. T., Knutsen, G., Goodfellow, M., Jones, A., Maldonado, L., Pathom-aree, W., Beil, W., Schneider, K., Keller, S. & Sussmuth, R. D. (2008). Proximicin A, B and C, novel aminofuran antibiotic and anticancer compounds isolated from marine strains of the actinomycete. *J. Antibiot.*, *61*, 158-163.

Freundlicha, J. S., Lalgondar, M., Wei, J. R., Swanson, S., Sorensen, E. J., Rubin, E. J. & Sacchettini, J. C. (2010). The Abyssomicin C family as in vitro inhibitors of Mycobacterium tuberculosis. *Tuberculosis (Edinb).*, *90*, 298-300.

Fu, P., Liu, P., Qu, H., Wang, Y., Chen, D., Wang, H., Li, J. & Zhu, W. (2011). α-Pyrones and diketopiperazine derivatives from the marine derived actinomycete *Nocardiopsis dassonvillei* HR10-5. *J. Nat. Prod.*, *74*, 2219-2223.

Fu, P., Yang, C., Wang, Y., Liu, P., Ma, Y., Xu, L., Su, M., Hong, K. & Zhu, W. (2012). Streptocarbazoles A and B, two novel indolocarbazoles from the marine-derived actinomycete strain *Streptomyces* sp. FMA. *Org. Lett.*, *14*, 2422-2425.

Galatenko, O. A. & Terekhova, L. P. (1990). Isolation of antibiotic-producing actinomycetes from soil samples exposed to UV-light. *Antibiot. I. Khimioter.*, *11*, 6-8.

Galatenko, O. A., Terekhova, L. P. & Bulina, T. I. (2004). A method for rapid search for actinomycete producers of antibiotics that are effective against methicillin-resistant *Staphylococcus aureus* strain. *Biotechnology., 4*, 17-23.

Goodfellow, M. & Haynes, J. A. (1984). Actinomycetes in marine sediments. In Biological, Biochemical, and Biomedical Aspects of Actinomycetes. Ortiz-Ortiz, L., Bojalil, L.F., and Yakoleff, V. (eds). New York, USA: Academic Press, pp. 453-472.

Goodfellow, M., Stach, J. E. M., Brown, R., Bonda, A. N. V., Jones, A. L., Mexson, J., Fiedler, H. P., Zucchi, T. D. & Bull, A. T. (2012). *Verrucosispora maris* sp. nov., a novel deep-sea actinomycete isolated from a marine sediment which produces abyssomicins. *Antonie van Leeuwenhoek., 101*, 185-193.

Goodfellow, M., Brown, R., Ahmed, L., Pathom-Aree, W., Bull, A. T., Jones, A. L., Stach, J. E., Zucchi, T. D., Zhang, L. & Wang, J. (2013). *Verrucosispora fiedleri* sp. nov., an actinomycete isolated from a fjord sediment which synthesizes proximicins. *Antonie van Leeuwenhoek., 103*, 493-502.

Gontang, E. A., Fenical, W. & Jensen, P. R. (2007). Phylogenetic diversity of Gram-positive bacteria cultured from marine sediments. *Appl. and Environ. Microbiol., 73*, 3272-3282.

Gonzalez, J. M. & Saiz-Jimenez, C. (2002). A fluorimetric method for the estimation of G+C mol% content in microorganisms by thermal denaturation temperature. *Environ. Microbiol., 4*, 770-773.

Gordon, R. E., Barnett, D. A., Handerhan, J. E. & Pang, C. H. N. (1974). *Nocardia coeliaca*, *Nocardia autotrophica*, and the nocardin strain. *Int. J Syst. Bacteriol., 24*, 54-63.

Gordon, R. E. & Mihm, J. M. (1962). Identification of *Nocardia caviae* (Erikson) nov. comb. *Ann. N. Y. Acad. Sci., 98*, 628-636.

Gulder, T. A. M. & Moore, B. S. (2010). Salinosporamide natural products: Potent 20S Proteasome inhibitors as promising cancer chemotherapeutics. *Angew. Chem. Int. Ed. Engl., 49*, 9346-9367.

Han, X. X., Cui, C. B., Gu, Q. Q., Zhu, W. M., Liu, H. B., Gu, J. Y. & Osada, H. (2005). ZHD-0501, a novel naturally occurring staurosporine analog from *Actinomadura* sp. 007. *Tetrahedron Lett., 46*, 6137-6140.

Hayakawa, M., Tamura, T., Iino, H. & Nonomura, H. (1991). Pollen-baiting and drying method for the highly selective isolation of Actinoplanes spp. from soil. *J. Ferment. Bioeng., 72*, 433-438.

Hohmann, C., Schneider, K., Bruntner, C., Brown, R., Jones, A. L., Goodfellow, M., Kramer, M., Imhoff, J. F., Nicholson, G., Fiedler, H. P. & Sussmuth, R. D. (2009). Albidopyrone, a new -pyrone-containing metabolite from marine-derived *Streptomyces* sp. NTK 227. *J. Antibiot.*, *62*, 75-79.

Hohmann, C., Schneider, K., Bruntner, C., Irran, E., Nicholson, G., Bull, A. T., Jones, A. L., Brown, R., StachJ, E. M., Goodfellow, M., Beil, W., Kramer, M., Imhoff, J. F., Sussmuth, R. D. & Fiedler, H. P. (2009). Caboxamycin, a new antibiotic of the benzoxazole family produced by the deep-sea strain *Streptomyces* sp. NTK 937. *J. Antibiot.*, *62*, 99-104.

Hsu, S. C. & Lockwood, J. L. (1975). Powdered chitin agar as a selective medium for enumeration of actinomycetes in water and soil. *Appl. Microbiol.*, *29*, 422-426.

Hu, Y., Martinez, E. D. & MacMillan, J. B. (2012). Anthraquinones from a Marine-Derived *Streptomyces spinoverrucosus*. *J. Nat. Prod.*, *75*, 1759-1764.

Huang, H., Yang, T., Ren, X., Liu, J., Song, Y., Sun, A., Ma, J., Wang, B., Zhang, Y., Huang, C., Zhang, C. & Ju, J. (2012). Cytotoxic angucycline class glycosides from the deep sea actinomycete *Streptomyces lusitanus* SCSIO LR32. *J. Nat. Prod.*, *75*, 202-208.

Huang, H., Yao, Y., ZHe, H., Yang, T., Ma, J., Tian, X., Li, Y., Huang, C., Chen, X., Li, W., Zhang, S., Zhang, C. & Ju, J. (2011). Antimalarial β-carboline and indolactam alkaloids from *Marinactinospora thermotolerans*, a deep sea isolate. *J. Nat. Prod.*, *74*, 2122-2127.

Hughes, C. C., Prieto-Davo, A., Jensen, P. R. & Fenical, W. (2008). The marinopyrroles, antibiotics of an unprecedented structure class from a marine *Streptomyces* sp. *Org. Lett.*, *10*, 629-631.

Igarashi, Y., Zhou, T., Sato, S., Matsumoto, T., Yu, L. & Oku, N. (2013). Akaeolide, a carbocyclic polyketide from marine-derived *Streptomyce*s. *Org. Lett.*, *15*, 5678-5681.

Ilan, E. Z., Torres, M. R., Prudhomme, J., Roch, K. L., Jensen, P. R. & Fenical, W. (2013). Farnesides A and B, sesquiterpenoid nucleoside ethers from a marine-derived *Streptomyces* sp., strain CNT-372 from Fiji. *J. Nat. Prod.*, *76*, 1815-1818.

Iwata, F., Sato, S., Mukai, T., Yamada, S., Takeo, J., Abe, A., Okita, T. & Kawahara, H. (2006). Lorneic Acids, trialkyl-substituted aromatic acids from a marine-derived actinomycete. *J. Nat. Prod.*, *72*, 2046-2048.

Jensen, P. R., Gontang, E., Mafnas, C., Mincer, T. J. & Fenical, W. (2005). Culturable marine actinomycete diversity from tropical Pacific Ocean sediments. *Environ. Microbiol.*, *7*, 1039-1048.

Jeong, S. Y., Shin, H. J., Kim, T. S., Lee, H. S., Park, S. K. & Kim, H. M. (2006). Streptokordin, a new cytotoxic compound of the methylpyridine class from a marine-derived *Streptomyces* sp. KORDI-3238. *J. Antibiot.*, *59*, 234-240.

Keller, S., Nicholson, G., Drahl, C., Sorensen, E., Fiedler, H. P. & Süssmuth, R. D. (2007). Abyssomicins G and H and atrop-abyssomicin C from the marine *Verrucosispora* strain AB-18-032. *J. Antibiot.*, *60*, 391-394.

Kim, S. B., Brown, R., Oldfield, C., Gilbert, S. C. & Goodfellow, M. (1999). *Gordonia desulfuricans* sp. nov., a benzothiophenedesulphurizing actinomycete. *Int. J. Syst. Bacteriol.*, *49*, 1845- 1851.

Kluge, A. G. & Farris, F. S. (1969). Quantitative phyletics and the evolution of anurans. *Syst. Zool.*, *18*, 1-32.

Kwon, H. C., Espindola, A. P. D. M., Park, J. S., Prieto-Davo, A., Rose, M., Jensen, P. R. & Fenical, W. (2010). Nitropyrrolins A-E, cytotoxic farnesyl--nitropyrroles from a marine-derived bacterium within the actinomycete family *Streptomycetaceae*. *J. Nat. Prod.*, *73*, 2047-2052.

Lavrova, N. V., Preobrazhenskaya, T. P. & Sveshnikova, M. A. (1972). Isolation of soil actinomycetes on selective media with rubomycin. *Antibiotics.*, *17*, 965-967.

Leutou, A. S., Yang, I., Kang, H., Seo, E. K., Nam, S. J. & Fenical, W. (2015). Nocarimidazoles A and B from a marine-derived actinomycete of the genus *Nocardiopsis*. *J. Nat. Prod.*, *78*, 2846-2849.

Li, Y. V., Terekhova, L. P. & Gapochka, M. G. (2002). Isolation of actinomycetes from soil using extremely high frequency radiation. *Microbiology.*, *71*, 105-108.

Li, F., Maskey, R. P., Qin, S., Sattler, I., Fiebig, H. H., Maier, A., Zeeck, A. & Laatsch, H. (2005). Chinikomycins A and B: isolation, structure elucidation, and biological activity of novel antibiotics from a marine *Streptomyces* sp. isolate M045. *J. Nat. Prod.*, *68*, 349-353.

Lu, Y., Xing, Y., Chen, C., Lu, J., Ma, Y. & Xi, T. (2012). Anthraquinone glycosides from marine *Streptomyces* sp. strain. *Phytochem. Lett.*, *5*, 459-462.

Macherla, V. R., Liu, J., Bellows, C., Teisan, S., Nicholson, B., Lam, K. S. & Potts, B. C. M. 2005. Glaciapyrroles A, B, and C, pyrrolosesquiterpenes from a *Streptomyces* sp. isolated from an alaskan marine sediment. *J. Nat. Prod.*, *68*, 780-783.

Macherla, V. R., Liu, J., Sunga, M., White, D. J., Grodberg, J., Teisan, S., Lam, K. S. & Potts, B. B. M. (2007). Lipoxazolidinones A, B, and C: Antibacterial 4-oxazolidinones from a marine actinomycete isolated from a Guam marine sediment. *J. Nat. Prod.*, *70*, 1454-1457.

Magarvey, N. A., Keller, J. M., Bernan, V., Dworkin, M. & Sherman, D. H. (2004). Isolation and characterization of novel marine-derived actinomycete taxa rich in bioactive metabolites. *Appl. Environ. Microbiol.*, *70*, 7520-7529.

Maldonado, L. A., Fenical, W., Jensen, P. R., Kauffman, C. A., Mincer, T. J., Ward, A. C., Bull, A. T. & Goodfellow, M. (2005a). *Salinispora arenicola* gen. nov., sp. nov. and *Salinispora tropica* sp. nov., obligate marine actinomycetes belonging to the family *Micromonosporaceae. Int. J. Syst. Evol. Microbiol.*, *55*, 1759-1766.

Maldonado, L. A., Stach, J. E. M., Pathom-aree, W., Ward, A. C., Bull, A. T. & Goodfellow, M. (2005b). Diversity of cultivable actinobacteria in geographically widespread marine sediments. *Antonie van Leeuwenhoek.*, *87*, 11-18.

Maldonado, L. A., Fragoso-Yanez, D., Perez-Garcı′, A., Rosellon-Druker, J. & Quintana, E. T. (2009). Actinobacterial diversity from marine sediments collected in Mexico. *Antonie van Leeuwenhoek.*, *95*, 111-120.

Maloney, K. N., MacMillan, J. B., Kauffman, C. A., Jensen, P. R., DiPasquale, A. G., Rheingold, A. L. & Fenical, W. (2009). *Org. Lett.*, *11*, 5422-5424.

Manam, R. R., Teisan, S., White, D. J., Nicholson, B., Grodberg, J., Neuteboom, S. T. C., Lam, K. S., Mosca, D. A., Lloyd, G. K. & Potts, B. C. M. (2005). Lajollamycin, a nitro-tetraene spiro--lactone--lactam antibiotic from the marine actinomycete *Streptomyces nodosus*. *J. Nat. Prod.*, *68*, 240-243.

Martin, G. D. A., Tan, L. T., Jensen, P. R., Dimayuga, R. E., Fairchild, C. R., Raventos-Suarez, C. & Fenical, W. (2007). Marmycins A and B, cytotoxic pentacyclic C-glycosides from a marine sediment-derived actinomycete related to the genus *Streptomyces*. *J. Nat. Prod.*, *70*, 1406-1409.

McArthur, K. A., Mitchell, S. S., Tsueng, G., Rheingold, A., White, D. J., Grodberg, J., Lam, K. S. & Potts, B. C. M. (2008). Lynamicins A-E, chlorinated bisindole pyrrole antibiotics from a novel marine actinomycete. *J. Nat. Prod.*, *71*, 1732-1737.

Mincer, T. J., Jensen, P. R., Kauffman, C. A. & Fenical, W. (2002). Widespread and persistent populations of a major new marine actinomycete taxon in ocean sediments. *Appl. Environ. Microbiol.*, *68*, 5005-5011.

Mincer, T. J., Fenical, W. & Jensen, P. R. (2005). Culture-dependent and culture-independent diversity within the obligate marine actinomycete genus *Salinispora. Appl. Environ. Microbiol., 71*, 7019-7028.

Minnikin, D. E., O'Donnell, A. G., Goodfellow, M., Alderson, G., Athalye, M., Schaal, A. & Parlett, J. H. (1984). An integrated procedure for the extraction of bacterial isoprenoid quinones and polar lipids. *J. Microbiol. Methods., 2*, 233-241.

Minnikin, D. E., Iwona, G., Hutchinson, G. & Caldicott, A. B. (1980). Thin-layer chromatography of methanolysates of mycolic acidcontaining bacteria. *J. Chromatogr., 188*, 221-233.

Mitchell, S. S., Nicholson, B., Teisan, S., Lam, S. K. & Potts, B. C. M. (2004). Aureoverticillactam, a novel 22-atom macrocyclic lactam from the marine actinomycete *Streptomyces aureoverticillatus. J. Nat. Prod., 67*, 1400-1402.

Motohashi, K., Takagi, M. & Shin-y, K. (2010). Tetrapeptides possessing a unique skeleton, JBIR-34 and JBIR-35, isolated from a sponge-derived actinomycete, *Streptomyces* sp. Sp080513GE-23. *J. Nat. Prod., 73*, 226-228.

Nam, S. J., Gaudencio, S. P., Kauffman, C. A., Jensen, P. R., Kondratyuk, T. P., Marler, L. E., Pezzuto, J. M. & Fenical, W. (2010). Fijiolides A and B, inhibitors of TNF--induced NFKB activation, from a marine-derived sediment bacterium of the genus *Nocardiopsis. J. Nat. Prod., 73*, 1080-1086.

Motohashi, K., Toda, T., Sue, M., Furihata, K., Shizuri, Y., Matsuo, Y., Kasai, H., Shin-ya, K., Takagi, M., Izumikawa, M., Horikawa, Y. & Seto, H. (2010). Isolation and structure elucidation of tumescenamides A and B, two peptides produced by Streptomyces tumescens YM23-260. *J. Antibiot., 63*, 549-552.

Nam, S. J., Kauffman, C. A., Jensen, P. R. & Fenical, W. (2011). Isolation and characterization of actinoramides A-C, highly modified peptides from a marine *Streptomyces* sp. *Tetrahedron., 67*, 6707-6712.

Nam, S. J., Kauffman, C. A., Paul, L. A., Jensen, P. R. & Fenical, W. (2013). Actinoranone, a cytotoxic meroterpenoid of unprecedented structure from a marine adapted S*treptomyces* sp. *Org. Lett., 15*, 5400-5403.

Nicolaou, K. C. & Harrison, S. T. (2007). Total synthesis of abyssomicin C, Atrop-abyssomicin C, and abyssomicin D: Implications for natural origins of Atrop-abyssomicin C. *J. Am. Chem. Soc., 129*, 429-440.

Nishimaru, T., Kondo, M., Takeshita, K., Takahashi, K., Ishihara, J. & Hatakeyama, S. (2014). Total synthesis of marinomycin A based on direct dimerization stretegy. *Angew. Chem. Int. Ed.*, *53*, 8459-8462.

Niu, S., Li, S., Chen, Y., Tian, X., Zhang, H., Zhang, G., Zhang, W., Yang, X., Zhang, S., Ju, J. & Zhang, C. (2011). Lobophorins E and F, new spirotetronate antibiotics from a South China Sea-derived *Streptomyces* sp. SCSIO 01127. *J. Antibiot.*, *64*, 711-716.

Nong, X. H., Zhang, X. Y., Xu, X. Y., Wang, J. & Qi, S. H. Nahuoic acids B−E, polyhydroxy polyketides from the marine-derived *Streptomyces* sp. SCSGAA 0027. *J. Nat. Prod.*, *79*, 141-148.

Oh, D. C., Gontang, E. A., Kauffman, C. A., Jensen, P. R. & Fenical, W. (2008). Salinipyrones and pacificanones, mixed-precursor polyketides from the marine actinomycete *Salinispora pacifica*. *J. Nat. Prod.*, *71*, 570-575.

Oh, D. C., Williams, P. G., Kauffman, C. A., Jensen, P. R. & Fenical, W. (2006). Cyanosporasides A and B, chloro- and cyano-cyclopenta[*a*]indene glycosides from the marine actinomycete "*Salinispora pacifica*". *Org. Lett.*, *8*, 1021-1024.

Palleroni, N. J. (1980). A chemotactic method for the isolation of Actinoplanaceae. *Arch. Microbiol.*, *128*, 53-55.

Paraskevi, N., Lampadariou, P. N., Mandalakis, M. & Tselepides, A. (2009). Phylogenetic diversity of sediment bacteria from the southern Cretan margin, Eastern Mediterranean Sea. *Syst. App. Microbiol.*, *32*, 17-26.

Phongsopitanun, W. C. Thawai, K., Suwanborirux, T., Kudo, M. & Ohkuma, Tanasupawat, S. (2014). *Streptomyces chumphonensis* sp. nov., isolated from marine sediments. *Int. J. Syst. Evol. Microbiol.*, *64*, 2605-2610.

Phongsopitanun, W., T., Kudo, M., Mori, K. Shiomi, P., Pittayakhjonwut, K. & Suwanborirux, Tanasupawat, S. (2015). *Micromonospora fluostatini* sp. nov., isolated from marine sediment. *Int. J. Syst. Evol. Microbiol.*, *65*, 4417-4423.

Polymenakou, P. N., Lampadariou, N., Mandalakis, M. & Tselepides, A. (2009). Phylogenetic diversity of sediment bacteria from the southern Cretan margin, Eastern Mediterranean Sea. *Syst. Appl. Microbiol.*, *32*, 17-26.

Prabahar, V., Dube, S., Reddy, G. S. N. & Shivaji, S. (2004). *Pseudonocardia antarctica* sp. nov. an actinomycetes from McMurdo Dry Valleys, Antarctica. *Sys. Appl Microbiol.*, *27*, 66-71.

Raju, R., Piggott, A. M., Quezada, M. & Capon, R. J. (2013). Nocardiopsins C and D and nocardiopyrone A: new polyketides from an Australian marine-derived *Nocardiopsis* sp. *Tetrahedron.*, *69*, 692-698.

Riedlnger, J., Reicke, A., Zahner, H., Krismer, B., Bull, A. T., Maldonado, L. A., Ward, A. C., Goodfellow, M., Bister, B., BIischoff, D., Sussmuth, R. & Fiedler, H. P. (2004). Abyssomicins, inhibitors of the para-aminobenzoic acid pathway produced by the marine *Verrucosispora* strain AB-18-032. *J. Antibiot.*, *57*, 271-279.

Rheims, H., Schumann, P., Rohde, M. & Stackebrandt, E. (1998). *Verrucosispora gifhornensis* gen. nov., sp. nov., a new member of the actinobacterial family *Micromonosporaceae. Int. J. Syst. Bacteriol.*, *48*, 1119-1127.

Saitou, N. & Nei, M. (1987). The neighbor-joining method: a new method for reconstructing phylogenetic trees. *Mol. Biol. Evol.*, *4*, 406-425.

Sato, S., Iwata, F., Yamada, S., Kawahara, H. & Katayama, M. (2011). Usabamycins A–C: New anthramycin-type analogues from a marine-derived actinomycete. *Bioorg. Med. Chem.*, *21*, 7099-7101.

Shin, H. J., Kim, T. S., Lee, H. S., Park, J. Y., Choi, I. K. & Kwon, H. J. (2008). Streptopyrrolidine, an angiogenesis inhibitor from a marine-derived *Streptomyces* sp. KORDI-3973. *Phytochem.*, *69*, 2363-2366.

Shirling, E. B. & Gottlieb, D. (1966). Methods for characterization of *Streptomyces* species. *Int. J. Syst. Bacteriol.*, *16*, 313-340.

Song, Y., Huang, H., Chen, Y., Ding, J., Zhang, Y., Sun, A., Zhang, W. & Ju, J. (2013). Cytotoxic and antibacterial marfuraquinocins from the deep South China Sea-derived *Streptomyces niveus* SCSIO 3406. *J. Nat. Prod.*, *76*, 2263-2268.

Soria-Mercado, I. E., Prieto-Davo, A., Jensen, P. R. & Fenical, W. (2005). Antibiotic terpenoid chloro-dihydroquinones from a new marine actinomycete. *J. Nat. Prod.*, *68*, 904-910.

Stach, J. E. M., Maldonado, L. A., Masson, D. G., Ward, A. C., Goodfellow, M. & Bull, A. T. (2003). Statistical approaches for estimating actinobacterial diversity in marine sediments. *Appl. Environ. Microbiol.*, *69*, 6189-6200.

Stackebrandt, E., Rainey, F. A. & Ward-Rainey, N. L. (1997). Proposal for a new hierarchic classification system, Actinobacteria classis nov. *Int. J. Syst. Bacteriol.*, *47*, 479-491.

Sun, P., Maloney, K. N., Nam, S. J., Haste, N. M., Raju, R., Aalbersberg, W., Jensen, P. R., VNizet, V., Hensler, M. E. & Fenical, W. (2011).

Fijimycins A–C, three antibacterial etamycin-class depsipeptides from a marine-derived *Streptomyces* sp. *Bioorg. Med Chem.*, *19*, 6557-6562.

Supong, K., Suriyachadkun, C., Tanasupawat, S., Suwanborirux, K., Pittayakhajonwut, P., Kudo, T. & Thawai, T. (2013). Micromonospora sediminicola sp. nov., isolated from marine sediment. *Int. J. Syst. Evol. Microbiol.*, *63*, 570-575.

Takada, K., Ninomiya, A., Naruse, M., Sun, Y., Miyazaki, M., Nogi, Y., Okada, S. & Matsunaga, S. (2013). Surugamides A−E, cyclic octapeptides with four Damino acid residues, from a marine *Streptomyces* sp.: LC−MS-aided inspection of partial hydrolysates for the distinction of D- and Lamino acid residues in the sequence. *J. Org. Chem.*, *78*, 6746-6750.

Tamura, K., Peterson, D., Peterson, N., Stecher, G., Nei, M. & Kumar, S. (2011). MEGA5: Molecular Evolutionary Genetics Analysis using Maximum Likelihood, Evolutionary Distance, and Maximum Parsimony Methods. *Mol. Biol. Evol.*, *28*, 2731-2739.

Terekhova, L. P., Galatenko, O. A., Alferova, I. V. & Preobragenskaya, T. P. (1991). Comparative estimation of some bacterial growth inhibitors as selective agents for isolation of soil actinomycetes. *Antibiotics Chemotherapy.*, *36*, 5-8.

Thompson, J. D., Gibson, T. J., Plewniak, F., Jeanmougin, F. & Higgins, D. G. (1997). The CLUSTAL_X windows interface: flexible strategies for multiple sequence alignment aided by quality analysis tools. *Nucleic Acids Res.*, *25*, 4876-4882.

Tian, X. P., Tang, S. K., Dong, J. D., Zhang, Y. Q., Xu, L. H., Zhang, S. & Li, W. J. (2009). *Marinactinospora thermotolerans* gen. nov., sp. nov., a marine actinomycete isolated from a sediment in the northern South China Sea. *Int. J. Syst. Evol. Microbiol.*, *59*, 948-952.

Tian, X. P., Zhi, X. Y., Qiu, Y. Q., Zhang, Y. Q., Tang, S. K., Xu, L. H., Zhang, S. & Li, W. J. (2009b). *Sciscionella marina* gen. nov., sp. nov., a marine actinomycete isolated from a sediment in the northern South China Sea. *Int. J. Syst. Evol. Microbiol.*, *59*, 222-228.

Um, S., Kim, Y. J., Kwon, H., Wen, H., Kim, S. H., Kwon, H. C., Park, S., Shin, J. & Oh, D. C. (2013). Sungsanpin, a lasso peptide from a deep-sea *Streptomycete*. *J. Nat. Prod.*, *76*, 873-879.

Williams, D. E., Dalisay, D. S., Li, F., Amphlett, J., Maneerat, W., Chavez, M. A. C., Wang, Y. A., Matainaho, T., Yu, W., Brown, P. J., Arrowsmith, C. H., Vedadi, M. & Andersen, R. J. (2013). *Org. Lett.*, *15*, 414-417.

Williams, D. E., Izard, F., Arnould, S., Dalisay, D. S., Tantapakul, C., Maneerat, W., Matainaho, T., Julien, E. & Andersen, R. J. (2016). Structures of Nahuoic acids B–E produced in culture by a *Streptomyces* sp. isolated from a marine sediment and evidence for the inhibition of the histone methyl transferase SETD8 in human cancer cells by nahuoic acid A. *J. Org. Chem.* DOI: 10.1021/acs.joc.5b02569.

Williams, P. G., Asolkar, R. N., Kondratyuk, T., Pezzuto, J. M., Jensen, P. R. & Fenical, W. (2007). Saliniketals A and B, bicyclic polyketides from the marine actinomycete *Salinispora arenicola*. *J. Nat. Prod.*, *70*, 83-88.

Williams, P. G., Buchanan, G. O., Feling, R. H., Kauffman, C. A., Jensen, P. R. & Fenical, W. (2005). New cytotoxic salinosporamides from the marine actinomycete *Salinispora tropica*. *J. Org. Chem.*, *70*, 6196-6203.

Williams, P. G., Miller, E. D., Asolkar, R. N., Jensen, P. R. & Fenical, W. (2007). Arenicolides A-C, 26-membered ring macrolides from the marine actinomycete *Salinispora arenicola*. *J. Org. Chem.*, *72*, 5025-5034.

Williams, S. T. & Cross, T. (1971). Actinomycetes. *Methods in Microbiology.*, *4*, 295-334. Edited by Booth, C. London: Academic Press.

Wu, S. J., Fots, S., Li, F., Qin, S., Kelter, G., Fiebig, H. H. & Laatsch, H. (2006). *N*-carboxamido-staurosporine and selina-4(14), 7(11)-diene-8, 9-diol, new metabolites from a marine *Streptomyces* sp. *J. Antibiot.*, *59*, 331-337.

Wu, C., Tan, Y., Gan, M., Wang, Y., Guan, Y., Hu, X., Zhou, H., Shang, X., You, X., Yang, Z. & Xiao, C. (2013). Identification of elaiophylin derivatives from the marine-derived actinomycete S*treptomyces* sp. 7145 using PCR-based screening. *J. Nat. Prod.*, *76*, 2153-2157.

Xi, L., Zhang, L., Ruan, J. & Huang Y. (2012). Description of *Verrucosispora qiuiae* sp. nov., isolated from mangrove swamp sediment, and emended description of the genus *Verrucosispora. Int. J. Syst. Evol. Microbiol.*, *62*, 1564-1569.

Xu, L. Y., Quan, X. S., Wang, C., Sheng, H. F., Zhou, G. X., Lin, B. R., Jiang, R. W. & Yao, X. S. (2011). Antimycins A19 and A20, two new antimycins produced by marine actinomycete *Streptomyces antibioticus* H74-18. *J. Antibiot.*, *64*, 661-665.

Xu, Y., He, J., Tian, X. P., Li, J., Yang, L. L., Xie, Q., Tang, S. K., Chen, Y-G., Zhang, S. & Li, W. J. (2012). *Streptomyces glycovorans* sp. nov., *Streptomyces xishensis* sp. nov. and *Streptomyces abyssalis* sp. nov., isolated from marine sediments. *Int. J. Sys. Evol. Microbiol.*, *62*, 2371-2377.

Yu, Z., Vodanovic-Jankovic, S., Ledeboer, N., Huang, S. X., Rajski, S. R., Kron, M. & Shen, B. (2011). Tirandamycins from *Streptomyces* sp. 17944 inhibiting the parasite *Brugia malayi* asparagine tRNA synthetase. *Org. Lett.*, *13*, 2034-2037.

Zhang, W., Liu, Z., Li, S., Lu, Y., Chen, Y., Zhang, H., Zhang, G., Zhu, Y., Zhang, G., Zhang, W., Liu, J. & Zhang, C. (2012). Fluostatins I−K from the south China sea-derived *Micromonospora rosaria* SCSIO N160. *J. Nat. Prod.*, *75*, 1937-1943.

Zhou, X., Huang, H., Chen, Y., Tan, J., Song, Y., Zou, J., Tian, X., Hua, Y. & Ju, J. (2012). Marthiapeptide A, an anti-infective and cytotoxic polythiazole cyclopeptide from a 60 L scale fermentation of the deep sea derived *Marinactinospora thermotolerans* SCSIO 00652. *J. Nat. Prod.*, *75*, 2251-2255.

In: Marine Sediments
Editor: Shirley Williams
ISBN: 978-1-63485-127-5

Chapter 2

EVALUATION OF MICRO-FABRIC NETWORK WITHIN MARINE SEDIMENTS BASED ON A ROCK MAGNETIC TECHNIQUE

***Yasuto Itoh*[1,*]*, Osamu Takano*[2]**
***and Machiko Tamaki*[3]**
[1]Graduate School of Science,
Osaka Prefecture University, Sakai, Osaka, Japan
[2]JAPEX Research Center,
Japan Petroleum Exploration Co., Ltd., Chiba, Japan
[3]Japan Oil Engineering Co., Ltd., Tokyo, Japan

ABSTRACT

Magnetic techniques that use anisotropy of magnetic susceptibility (AMS) act as a proxy of preferred permeable orientation in basin-filling sediments, when it is applied on samples impregnated with a magnetic suspension. The unique method for quantifying heterogeneity in rocks is reviewed and its value for reconstruction of the preferred direction of pore fluid flow is reassessed critically. The authors also present results of their experiments, which dealt with secondary fracture networks developed in tight sandstones burying a foreland basin on an arc-arc collision zone. Directional analysis of AMS ellipsoid implies tectonic

* Correspondence: yasutokov@yahoo.co.jp.

control on rupture development under strong trans-compressive regime. Micro-focus three-dimensional density imaging of test pieces has shown a substantial variation in pore fabric reflecting inhomogeneous impregnation of magnetic fluid within rocks.

1. INTRODUCTION

Micro-fabric of marine sediments is a versatile petrophysical information reflecting clast alignment for the reconstruction of sediment transport mechanism, direction of paleoflow and preferred permeable orientation of formation fluids since basins remote from hinterlands are filled with fine deposits lacking in visible manifestation of sedimentary structure. Hence many studies have explored standardized procedures for acquisition of grain fabric through conventional microscopy, which was summarized as an extensive review (Baas et al., 2007).

Baas et al., (2007) also evaluated a brand-new method utilizing magnetic techniques based on anisotropy of magnetic susceptibility (AMS) as a proxy for grain orientation, which benefits greatly from being able to measure a large number of grains in a three-dimensional space, in a short amount of time and with a lower sensitivity to user bias. The magnetic technique applied on samples impregnated with liquid containing suspension of ferromagnetic powder was originally developed for the purpose of oil exploration by Hailwood and his colleague (e.g., Hailwood et al., 1999) on the theoretical background of Pfleiderer and Halls (1994).

Because the rock magnetic method was first applied to continental setting where high porous sediments are widely distributed, previous studies tend to concentrate on evaluation of fabric of intact detrital grains and their interstitial pore geometry, omitting tectonic control on fracture generation and enhancement of effective permeability. The authors' preliminary study (Itoh et al., 2014a) on collision-related turbidite samples showed wider variety in micro-fabric reflecting secondary fracture network under strong tectonic stress. In this paper, we aim at verification of AMS usability as a textural indicator of rocks, with special emphasis on origin of fluid pathway within tight turbidite sandstones burying a foreland basin.

2. Methodology: A Review

2.1. Anisotropy of Magnetic Susceptibility

To measure the initial magnetic susceptibility, a rock sample is placed in a low-intensity magnetic field of strength H, and the intensity of the induced magnetization (J) is measured for different orientations of the field. The induced magnetization is related to the field strength through the magnetic susceptibility (K) of the sample, where:

J=K

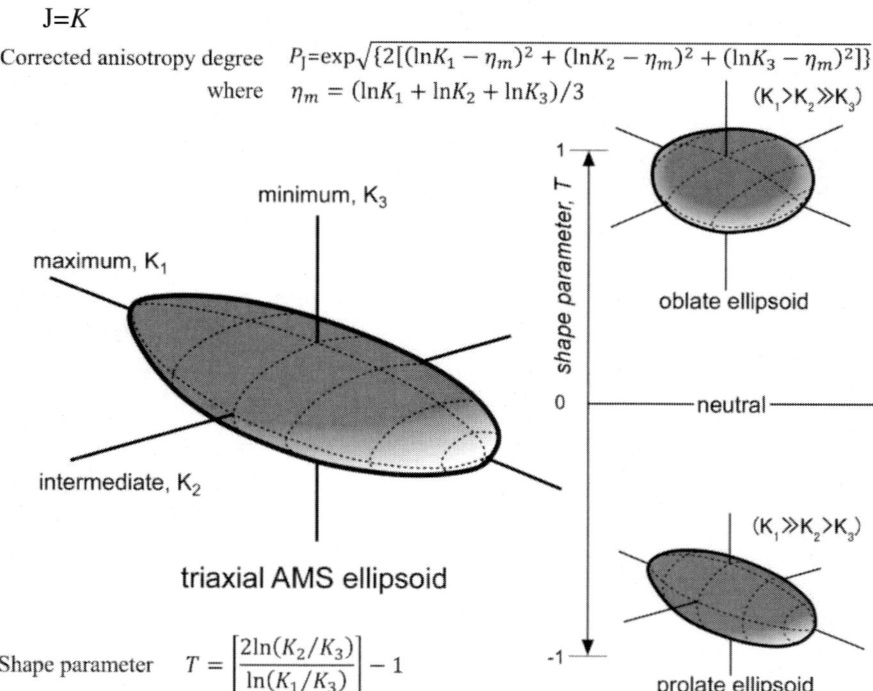

Figure 1. The susceptibility ellipsoids. Significant parameters of anisotropy of magnetic susceptibility (AMS), such as P_J and T, are calculated based on orthogonal principal susceptibilities (K_1, K_2, K_3).

The magnetic susceptibility of a sample is the summation of the susceptibilities of all mineral species within rock samples, and magnetic susceptibility is rarely identical in all directions of measurement; it more or less is an anisotropic parameter. The degree of anisotropy of magnetic susceptibility (AMS) is dependent primarily on mineral type, but even for a single type of mineral species the magnitude and anisotropy of magnetic susceptibility may considerably vary with many physicochemical properties (Tarling and Hrouda, 1993). Figure 1 presents significant parameters of AMS with typical types of susceptibility ellipsoids.

2.2. Conventional Method

As for common sedimentary rocks with low content of ferromagnetic minerals, AMS fabric generally reflects alignment of iron-rich silicates such as biotite and amphibole, of which orientations are bound to sedimentary structure. Based on azimuth of untilted AMS principal axes of fine sediments, Itoh et al., (2006) argued that the AMS is controlled by shape anisotropy of minerals laid on bedding plane.

Most igneous rocks are magnetically isotropic in primary state. Secondary structure may, however, cause detectable AMS trend. Based on intensive rock magnetic experiments, Itoh and Amano (2004) found that an enhanced AMS trend near a major fault cutting a granitoid pluton is originated from fine-grained authigenic magnetite grains precipitated on fracture surface.

Although such conventional methods are effective in case the possible AMS-carrying minerals are rather restricted, anisotropy of individual particles is brought about by the magnetocrystalline anisotropy or by the shape anisotropy, of which contributions vary with mineral species. Therefore more direct analytical method is necessary for quantitative description of micro-fabric of rocks.

2.3. Ferrofluid Method

Figure 2 shows a generalized scheme of analytical method utilizing liquid containing suspension of magnetic powder (hereafter referred to as ferrofluid). Since ultrafine ferromagnetic minerals in ferrofluid (mainly magnetite) are under superparamagnetic limit, coercive force (H_C) is actually negligible. As a result, maximum axis of AMS (K_1) of a rock sample impregnated with ferrofluid reflects elongate azimuth of pore network and most permeable direction.

Since the pioneering work of Hailwood, many researchers have applied the attractive method to various research areas. Nabawy et al., (2009) evaluated permeability and magnetic pore fabrics of an aquifer, and stated that pore fabrics of most formations in the Tushka Basin, Egypt are closely linked to paleocurrent directions with minor exception showing structural control (affinity for the main fault trends around the study area).

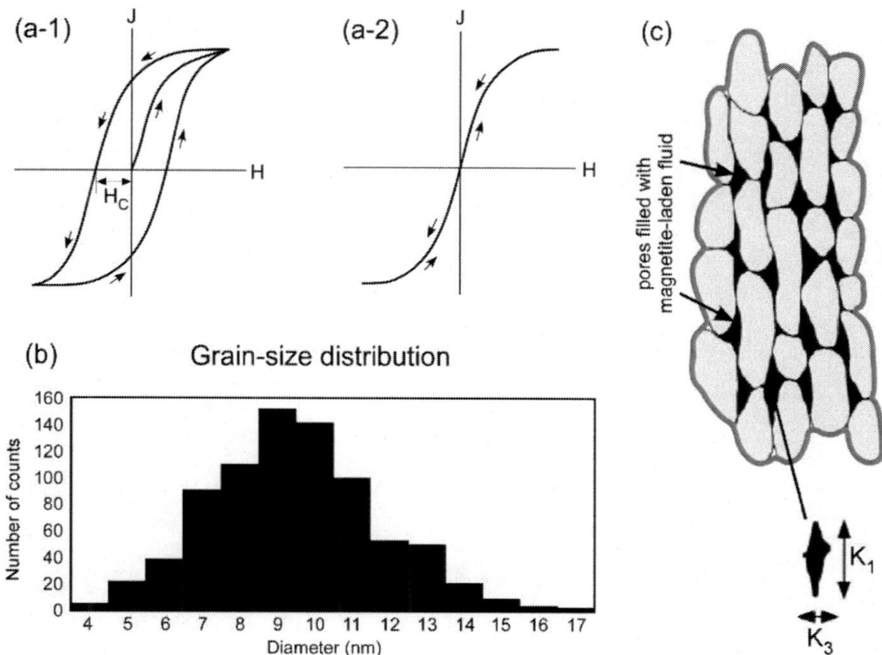

Figure 2. Generalized basis of analytical method utilizing ferrofluid. (a) Comparison of magnetic hysteresis between conventional ferromagnetic body (a-1) and ferrofluid (a-2). Ultrafine ferromagnetic minerals (mainly magnetite) are under superparamagnetic limit, and Hc is negligible. (b) Grain size distribution of a ferrofluid (provided by FerroTec Co., Ltd.). (c) Schematic view of a rock sample impregnated with ferrofluid (darkened parts).

Almqvist et al. (2011) utilized AMS-derived pore shape geometry for prediction of elastic properties for porous and anisotropic synthetic aggregates. It is noted that they adopted X-ray micro-tomography density contrast imaging, where attenuation of X-rays is related to the density contrast of the material, of dry and ferrofluid-processed specimens, and visually evaluated completeness of fluid saturation. At the end of this paper, we attempt to visualize various patterns of impregnation using similar apparatus.

Reflecting highly porous nature of samples in continental setting, the above studies used a vacuum chamber to impregnate ferrofluid. Although Baas et al., (2007) made a comment on high pressure treatment for complete saturation with ferrofluid, detailed experimental condition was not shown.

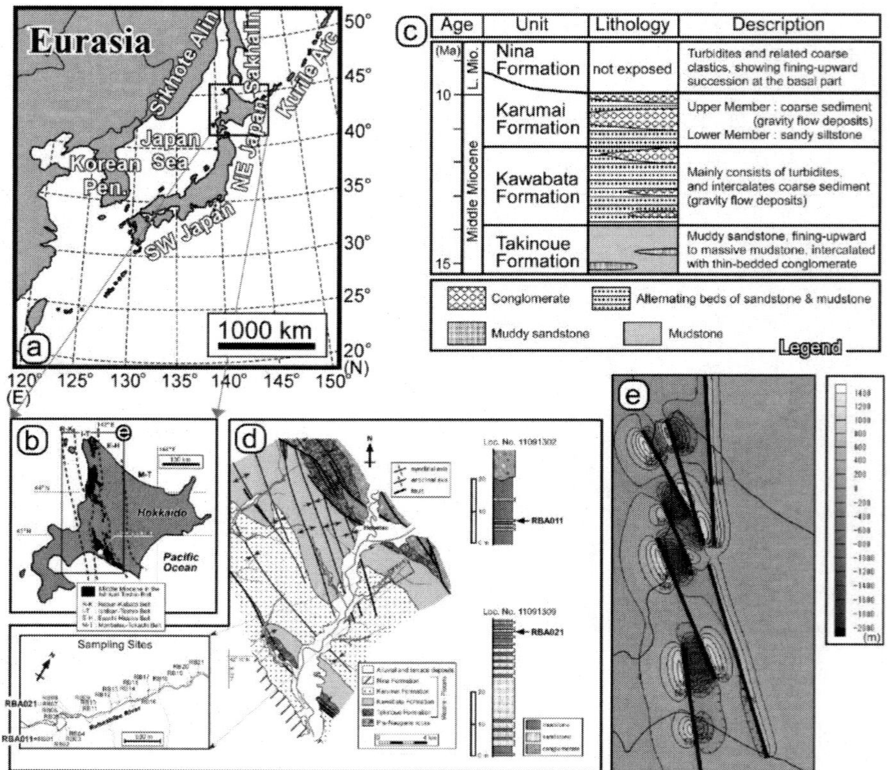

Figure 3. Geological background of the case study area. (a) Regional index. (b) Cenozoic tectonic context of central Hokkaido. (c) Neogene stratigraphy of the study area. (d) Geology of the study area (simplified from Kawakami et al., 1999), and locations of rock magnetic samples (modified from Itoh et al., 2014a). (e) Dislocation modeling of the Kawabata sedimentary basin (modified from Itoh et al., 2014b). Mapped area is shown by an envelope in (b).

3. Case Study

3.1. Geological Background

Sedimentary basins on convergent margins are rapidly filled by voluminous clastics, part of which is volcaniclastic reflecting active volcanism, then deformed and exhumed under compressive stress provoked by various tectonic events. On such condition, tight sedimentary rocks suffering diagenesis are less porous and instead studded by numerous secondary fractures. One presumable theory is that micro-fabrics of rocks are different from those in stable continental setting.

Our focus is set on a turbidite sequence, Kawabata Formation, widespread in central Hokkaido, which deposited during a middle Miocene arc-arc collision event. Figure 3 presents geological background of the case study area. Hokkaido is located on the northeastern margin of Eurasia, and has suffered various tectonic events through the Cenozoic. Reflecting backarc spreading and subsequent collision events, Neogene sequence of the study area shows a transgression-regression cycle (see Figure 3c). Recently a numerical modeling was executed on the Kawabata sedimentary basin (Itoh et al., 2014b) and clarified that transpressional regime followed by strong compression on the NNW-SSE regional fault zone is essential to restore the basin configuration (Figure 3e).

3.2. Previous Studies

The authors executed preliminary rock magnetic studies on the Kawabata Formation (Itoh et al., 2013, 2014a). Rock magnetic samples were taken along the Rubeshibe river in Hobetsu district in central Hokkaido, where Kawakami et al., (1999) reported detailed stratigraphic and structural data (see Figure 3d).

Itoh et al., (2013) collected samples of the Kawabata Formation with a battery-powered electric drill at 21 sites along the Rubeshibe route. The bedding attitudes were measured on outcrops to compensate for tectonic tilting later. Between seven and sixteen independently oriented cores 25 mm in diameter were obtained at each site using a magnetic compass. Cylindrical specimens 22 mm in length were cut from each core and the natural remanent magnetization (NRM) of each specimen was measured using a cryogenic magnetometer (model 760-R SRM, 2-G Enterprises). Low-field magnetic susceptibility was measured on a Bartington MS2 susceptibility meter, and the AMS was measured using an AGICO KappaBridge KLY-3 S magnetic susceptibility meter.

Their sedimentological results are shown in Figure 4. AMS fabrics of most raw samples of the Kawabata Formation are highly oblate, as shown by positive T parameters near unity. This fabric is essentially confined to the bedding plane under gravitational force, and the authors considered the fabric as being governed simply by the shape anisotropy of paramagnetic minerals, i.e., alignments of elongate or platy grains such as amphibole or mica based on hysteresis study indicative of a negligible amount of ferromagnetic material.

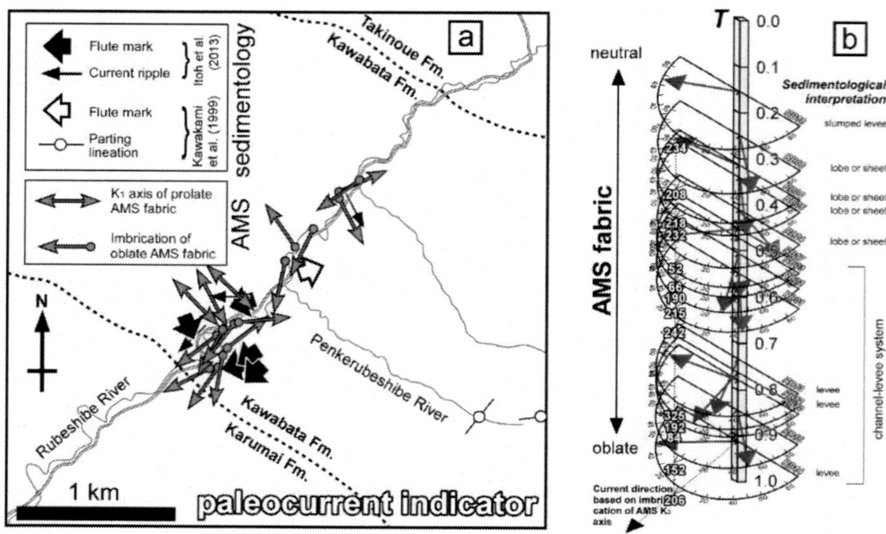

Figure 4. Summary of previous sedimentological analyses modified from Itoh et al., (2013). (a) Paleocurrent map of the Kawabata Formation around the Rubeshibe River route. Formation boundaries are after Kawakami et al., (1999). (b) AMS paleocurrent indicators of the Kawabata Formation. Directions of K_1 (gray arrows) are shown as acute angles from the dotted baseline of K_3 axis imbrication. Downcurrent orientations based on imbrication data are depicted as outlined numbers on the baseline. Vertical positions of the data represent degree of AMS oblateness shown by the T parameter. Samples with negative T values are excluded from the diagram because such cases have a large scatter in the K_3 directions.

Notably, the imbrication of the oblate AMS fabric matches visible sedimentary structures (Figure 4a), suggesting that AMS data can serve to indicate paleocurrents after the contributors to the magnetic fabric have been identified. It is also indicated that K_1 of prolate samples (with negative T parameters) tend to align perpendicular to the paleocurrent direction, implying that elongate grains roll on the sediment surface.

Figure 4b shows a series of paleocurrent indicators identified in the Kawabata Formation as a function of the AMS shape parameter (T). The intensity of alignment forcing inferred from AMS data is closely related to sedimentary facies (shown on the right in the figure) determined by field observation. For example, weak hydrodynamic forcing corresponds to fine rhythmically alternating facies in channel-levee systems. Thus, the sedimentological context of muddy sediments' AMS fabric can be interpreted in the light of sandy sediments' facies analysis.

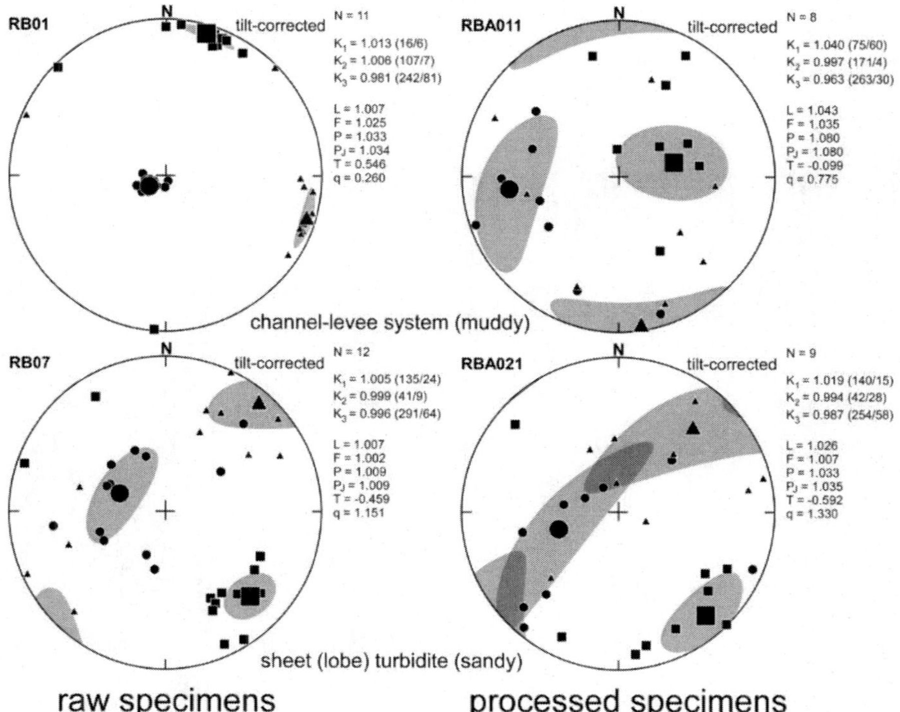

Figure 5. Tilt-corrected AMS fabric for the Kawabata Formation samples in raw (left) and ferrofluid-soaked (right) states. All the data are plotted on the lower hemisphere. Square, triangular and circular symbols represent orthogonal maximum (K_1), intermediate (K_2), and minimum (K_3) AMS principal axes, respectively, and larger symbols show their mean directions. Shaded areas are 95% confidence limits of Bingham statistics. Anisotropy parameters posted on the equal-area diagrams are calculated based on Tarling and Hrouda (1993). Generally, ferrofluid impregnation results in enhanced anisotropy degree. In site RB01 (upper), processed specimen (RBA011) shows prolate fabric and quite different spatial arrangement of principal axes from the raw data, whereas site RB07 (lower) is characterized by similar AMS trend after ferrofluid treatment (RBA021).

Based on the sedimentological discussion above, Itoh et al., (2014a) executed ferrofluid experiments on selected Kawabata samples in the Rubeshibe route. Hand samples oriented using a magnetic compass were taken from two sites, RBA011 (muddy channel and levee turbidite) from RB01 and

RBA021 (sandy sheet turbidite) from RB07. They were cut into cubic specimens with an approximate volume of 4 cm^3. Their permeability measured by a Pressure Decayed Profile Permeameter ranges 0.014~0.030 md for RBA011 and 0.053~0.151 md for RBA021, respectively. After evacuation for one day, all the samples were soaked in water-based ferrofluid (MSG W10 with saturation magnetization of 185 Gauss; provided by FerroTec Co., Ltd.) contained in a pressure vessel and impregnated under 5 MPa for 30 days.

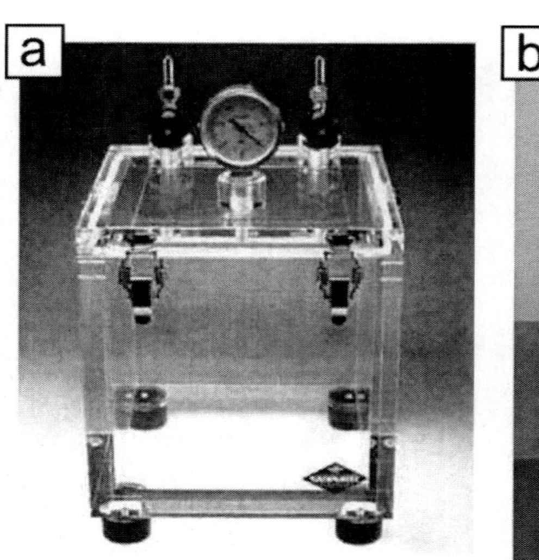

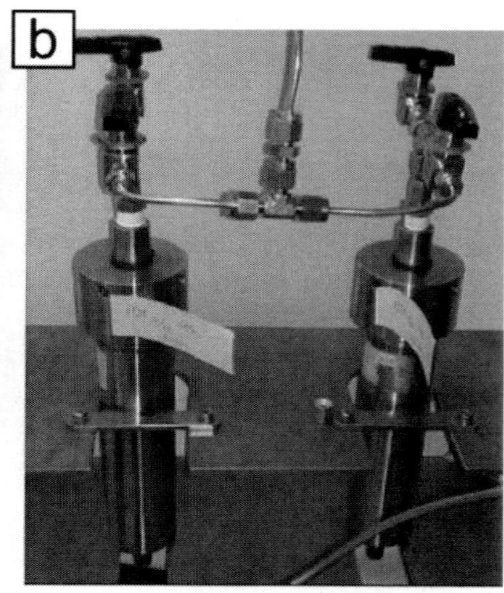

Figure 6. Experimental apparatuses to impregnate ferrofluid. (a) Vacuum chamber for low-pressure treatment. (b) Pressure vessel for high-pressure treatment.

Impregnated samples were washed with purified water, dried and then contained in plastic capsules. Their AMS data were measured using an AGICO KappaBridge KLY-3 S magnetic susceptibility meter, and much more than one digit larger bulk susceptibility indicates successful impregnation of ferrofluid. Measured AMS parameters were compared with those for raw samples in the same sites reported by Itoh et al., (2013). Figure 5 presents tilt-corrected AMS fabrics for the Kawabata Formation samples before and after ferrofluid treatment. Generally speaking, ferrofluid impregnation results in enhanced anisotropy degree (P_J). In site RB01, processed specimen (RBA011) shows prolate fabric and quite different spatial arrangement of principal axes from the raw data, whereas site RB07 is characterized by similar AMS trend after ferrofluid treatment (RBA021). They attributed such a variety in magnetic fabric to difference in microscopic sedimentary structure.

3.3. Experimental Scheme

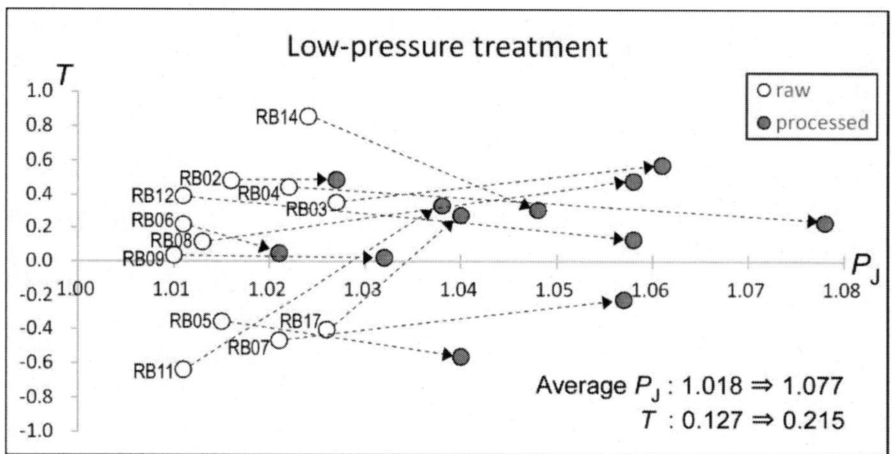

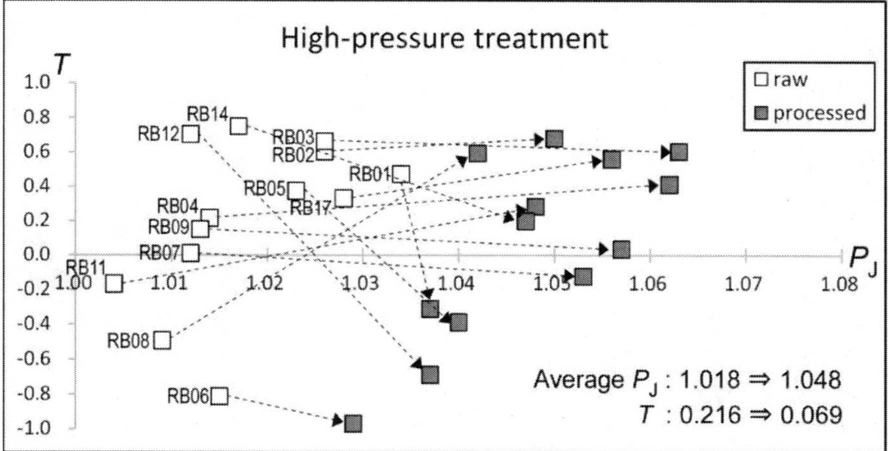

Figure 7. Magnitudes of magnetic fabrics in raw samples (open symbol) and ferrofluid-impregnated samples (solid symbol) for 13 sites of the Kawabata Formation. As for the low-pressure treatment, site RB01 is omitted because its processed P_J (1.448) is out of the range of transverse axis. Average P_J and T are arithmetic means of specimen data.

In the present study, the authors prepared two specimens (4 cm^3 oriented cubes cut from 25 mm diameter cores originally taken by Itoh et al., 2013) from 13 sites of the Kawabata Formation to observe lithofacies effect on the AMS fabrics. Using the water-based ferrofluid (MSG W10), one was impregnated in a vacuum chamber (Figure 6a) for 30 days, the other was impregnated in a pressure vessel (Figure 6b) under 5 MPa for 30 days. They were washed with purified water, dried and then contained in plastic capsules. Their AMS data were measured using an AGICO KappaBridge KLY-3 S magnetic susceptibility meter.

4. Results

Bulk susceptibility of the processed samples were one to two digits larger than raw state indicating successful impregnation of ferrofluid. As shown in Figure 7, magnitudes of magnetic fabrics in ferrofluid-impregnated samples (solid symbol) are much greater than raw samples (open symbol) for 13 sites of the Kawabata Formation. Low-pressure treatment did not so much affect shape parameter T, whereas some specimens show clear prolate fabric after high-pressure treatment. Thus we investigate a directional trend of AMS axes in the next section.

5. Discussion

5.1. AMS Fabric

Figure 8 delineates tilt-corrected AMS axes for 13 sites of the Kawabata Formation samples in raw (left) and ferrofluid-soaked (right) states. Generally, ferrofluid impregnation results in enhanced anisotropy degree. In case of low-pressure treatment (upper), AMS trend is more or less similar before and after ferrofluid experiment, whereas high-pressure treatment (lower) results in considerable decrease in T parameter although spatial arrangement of principal axes remained unchanged. This implies that pressurized ferrofluid was impregnated via a pathway which is impermeable under atmospheric pressure.

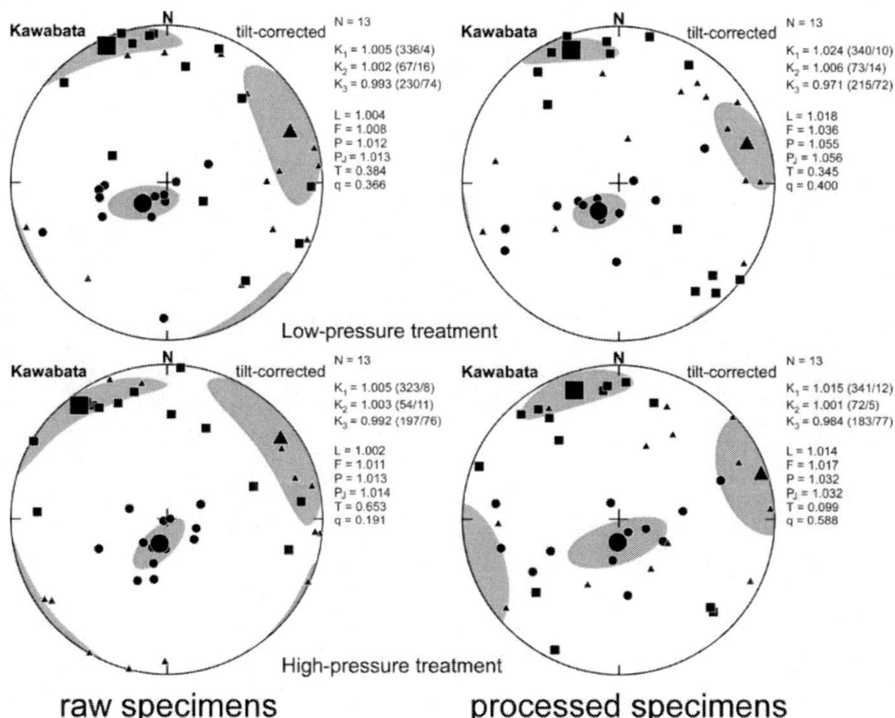

Figure 8. Tilt-corrected AMS fabric for 13 sites of the Kawabata Formation samples in raw (left) and ferrofluid-soaked (right) states. All the data are plotted on the lower hemisphere. Square, triangular and circular symbols represent orthogonal maximum (K_1), intermediate (K_2), and minimum (K_3) AMS principal axes, respectively, and larger symbols show their mean directions. Shaded areas are 95% confidence limits of Bingham statistics. Anisotropy parameters posted on the equal-area diagrams are calculated based on Tarling and Hrouda (1993). Generally, ferrofluid impregnation results in enhanced anisotropy degree. In case of low-pressure treatment (upper), AMS trend is more or less similar before and after ferrofluid experiment, whereas high-pressure treatment (lower) results in considerable decrease in T parameter. It is noted that azimuth of the K_1 axis, which is the most permeable direction, is coincident with regional fault system in central Hokkaido (see Figure 3).

5.2. Tectonic Implication

It is noted that azimuth of the K_1 axis after the high-pressure treatment, which is the most permeable direction in subsurface condition, is coincident with regional fault system in central Hokkaido (see Figure 3). Not only in the intensive collision event during the Kawabata stage, the NNW-SSE fault system has been intermittently activated with dextral slips throughout the Cenozoic era (Kusumoto et al., 2013). Thus the AMS data with ferrofluid treatment may delineate an invisible weakness in rocks in quantitative way.

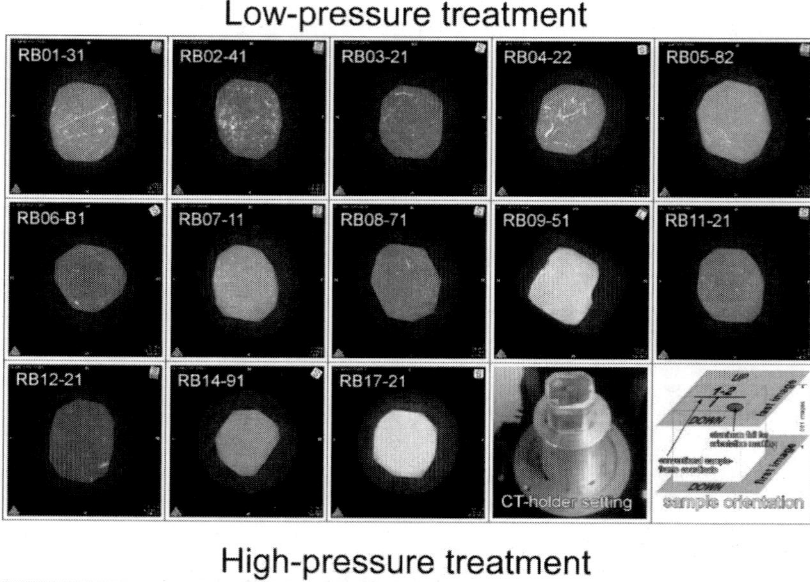

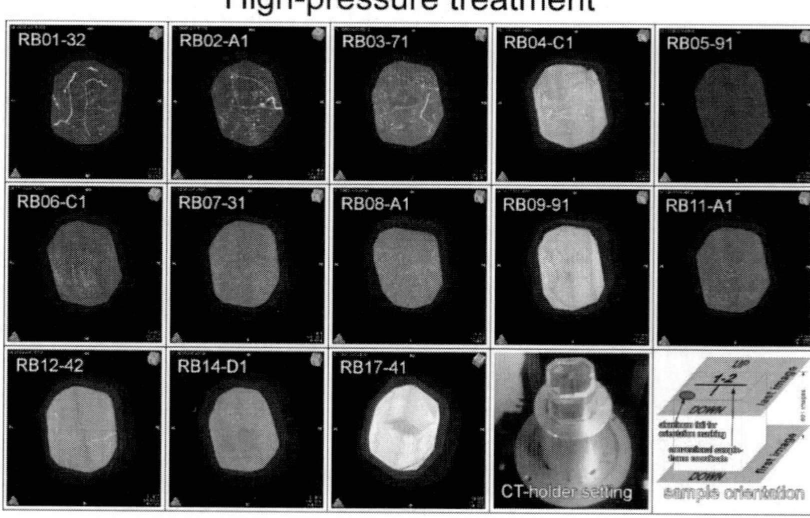

Figure 9. Three-dimensional maximum intensity projection (MIP) images of ferrofluid-processed specimens generated by OsiriX MD. Original image sequences (601 images per sample) were acquired using HMX225-ACTIS+3 Micro-Focus X-Ray CT Scanner at Center for Advanced Marine Core Research, Kochi University with 30 μm spatial resolution.

Conclusion

Well-organized magnetic experiments utilizing ferrofluid revealed wide variety in microscopic fabric of sedimentary rocks. Visualized spatial distribution of permeable pore spaces in rocks shows considerable diversity reflecting experimental methods (low or high pressure) and variety in lithologic facies. Figure 9 presents three-dimensional maximum intensity projection (MIP) images of ferrofluid-processed specimens generated by OsiriX MD. It is obvious that efficiency of impregnation differs even in the same site reflecting small-scale structural disturbance, some amount of which should be owing to bioturbation that is delineated by sinuous pathway of the dense fluid. A series of movies (13 files in .mov format each for low- and high-pressure treatments) of micro-focus X-ray CT scanning images (601 per sample at 30 μm spatial resolution) are available at OPERA: Osaka Prefecture University Education and Research Archives (http://hdl.handle.net/10466/14732).

Acknowledgments

The authors are grateful to N. Ishikawa for the use of the rock-magnetic laboratory at Kyoto University and for thoughtful suggestions in the course of the magnetic analyses. Thanks are also due to Y. Yamamoto and support staffs for their kind backup during experiments at Center for Advanced Marine Core Research, Kochi University. We thank M. Nakano and EOR Group staffs of JAPEX Research Center for their kind collaboration and superb operation during preparation of impregnated samples. This work was supported in part by the Grant-in-aid (No. 25400456) from Japanese Ministry of Education, Culture, Sports, Science and Technology (MEXT).

References

Almqvist, B. S. G., Mainprice, D., Madonna, C., Burlini, L., & Hirt, A. M. (2011). Application of differential effective medium, magnetic pore fabric analysis, and X-ray microtomography to calculate elastic properties of porous and anisotropic rock aggregates. *Journal of Geophysical Research*, 116, B01204, doi:10.1029/2010JB007750.

Baas, J. H., Hailwood, E. A., McCaffrey, W. D., Kay, M., & Jones, R. (2007). Directional petrological characterisation of deep-marine sandstones using grain fabric and permeability anisotropy: Methodologies, theory, application and suggestions for integration. *Earth-Science Reviews*, 82, 101-142.

Hailwood, E. A., Bowen, D., Ding, F., Corbett, P. W. M., & Whattler, P. (1999). Characterizing pore fabrics in sediments by anisotropy of magnetic susceptibility analyses, In D. H. Tarling and P. Turner (Eds.), Palaeomagnetism and Diagenesis in Sediments. *Geological Society, London, Special Publications*, 151, 125-126.

Itoh, Y., & Amano, K. (2004). Progressive segmentation and systematic block rotation within a plutonic body: palaeomagnetism of the Cretaceous Kurihashi granodiorite in northeast Japan. *Geophysical Journal International*, 157, 128-140.

Itoh, Y., Amano, K., & Kumazaki, N. (2006). Integrated description of deformation modes in a sedimentary basin: A case study around a shallow drilling site in the Mizunami area, eastern part of southwest Japan. *Island Arc*, 15, 165-177.

Itoh, Y., Tamaki, M., & Takano, O. (2013). Rock magnetic properties of sedimentary rocks in central Hokkaido - insights into sedimentary and tectonic processes on an active margin. In Y. Itoh (Ed.), Mechanism of Sedimentary Basin Formation - Multidisciplinary Approach on Active Plate Margins. Rijeka (Croatia): InTech. http://dx.doi.org/10.5772/56650.

Itoh, Y., Takano, O., & Tamaki, M. (2014a). Quantitative analysis of pore network in rocks utilizing ferrofluids: Application of rock magnetic data for oil exploration. *Journal of the Japanese Association for Petroleum Technology*, 79, 339-348.

Itoh, Y., Takano, O., Kusumoto, S., & Tamaki, M. (2014b). Mechanism of long-standing Cenozoic basin formation in central Hokkaido: an integrated basin study on an oblique convergent margin. *Progress in Earth and Planetary Science*, 1:6. doi:10.1186/2197-4284-1-6.

Kawakami, G., Yoshida, K., & Usuki, T. (1999). Preliminary study for the Middle Miocene Kawabata Formation, Hobetsu district, central Hokkaido, Japan: special reference to the sedimentary system and the provenance. *Journal of the Geological Society of Japan*, 105, 673-686.

Kusumoto, S., Itoh, Y., Takano, O., & Tamaki, M. (2013). Numerical modeling of sedimentary basin formation at the termination of lateral faults in a tectonic region where fault propagation has occurred. In Y. Itoh (Ed.), Mechanism of Sedimentary Basin Formation - Multidisciplinary

Approach on Active Plate Margins. Rijeka (Croatia): InTech. http://dx.doi.org/10.5772/56558.

Nabawy, B. S., Rochette, P., & Geraud, Y. (2009). Petrophysical and magnetic pore network anisotropy of some cretaceous sandstone from Tushka Basin, Egypt. *Geophysical Journal International*, 177, 43-61.

Pfleiderer, S., & Halls, H. C. (1994). Magnetic pore fabric analysis: a rapid method for estimating permeability anisotropy. *Geophysical Journal International*, 116, 39-45.

Tarling, D. H., & Hrouda, F. (1993). The Magnetic Anisotropy of Rocks. London: Chapman & Hall, 217pp.

In: Marine Sediments
Editor: Shirley Williams
ISBN: 978-1-63485-127-5

Chapter 3

ALUMINIUM IMPACT ON THE GROWTH OF BENTHIC DIATOM

***L. Leleyter*[1,*], *F. Baraud*[1], *O. Gil*[1], *S. Gouali*[1], *M. Lemoine* [1] *and F. Orvain*[2]**

[1]Université de Caen Basse-Normandie -
Unité de Recherche Aliments Bioprocédés Toxicologie Environnements (UR ABTE), Caen cedex, France

[2]Université de Caen Basse-Normandie - Biologie des Organismes Aquatiques et Ecosystèmes (UMR BOREA), France

ABSTRACT

A laboratory experiment was performed, in tidal artificial conditions, with natural defaunated sediments that were artificially contaminated with aluminium salts. Benthic diatoms were cultivated on these contaminated sediments in a tidal mesocosm with natural seawater, in order to assess the aluminium impact on the growth of benthic diatom. The sediments were also leached by HCl (1M) to estimate the available/mobile part of the aluminium in the sediment.

This experiment highlighted that Al-contamination sediments had a strong effect on diatom communities growing at sediments surface. Drastic negative effect is perceptible beyond 10 mg of aluminium salts added by kg of natural sediment (which is a low contamination stress

[*] Corresponding author: lydia.reinert@unicaen.fr.

with regard to the Al initial content). The 1M HCl-extraction, widely used in the literature for metals mobility evaluation in marine sediments seems to underestimate aluminium availabilities for phytobenthic communities.

Keywords: marine sediment, HCl extraction, aluminium, phytobentho

1. INTRODUCTION

Heavy metals contamination in sediments is often assessed by the determination of their total contents and comparison with either established national guidelines or some background reference values (Covelli and Fantolan, 1997; Giancoli Barreto et al., 2004; Leleyter and Baraud, 2005; Dubrulle et al., 2007; Meybeck et al., 2007; Chand and Prasad, 2013). However metal concentrations derived from these total digests do not necessarily provide biologically meaningful data. The alternative is to use a partial extraction that only targets the labile mineral phases, as these are most likely to exert an influence on biota (Tam et al., 1989; Scouller et al., 2006). Thus, other studies focused on the potential mobility of these heavy metals in the sediments, estimated by either simple or sequential extraction procedures (Tessier et al., 1979; Leleyter and Probst, 1999; Sutherland, 2002; Rousseau et al., 2009; Devesa-Rey et al., 2010; Leleyter et al., 2012; Roussiez et al., 2013; El Azzi et al., 2013; Alvarenga et al., 2014; Islam et al., 2015; Hamdoun et al., 2015a and b).

For marine sediments, HCl (1M) extraction is recommended by many authors (Doherty et al., 2000; Snape et al., 2004; Burton et al., 2005; Scouler et al., 2006; Larner et al., 2008; Leleyter et al., 2012; Pena-Icart et al., 2014; Hamdoun et al., 2015a and b), as HCl is assumed to extract heavy metals thanks to its acidic properties combined with the chelatant property of Cl^-. For a comprehensive environmental risk assessment, both approaches (total and labile contents) are generally complementary in order to fully characterize the degree of sediments contamination.

Natural aluminium (Al) is the third most abundant element in the earth crust (around 8%, Allègre and Michard, 1973). Aluminium is known as a refractory element, typically not diagenetically labile. Moreover, anthropological activities (aluminium industry, water treatments, steel protection with aluminium sacrificial anodes, ...) release aluminium in the environment. As aluminum is more stable in solid than in aqueous phase, the anthropological aluminum poured in marine waters tends to be sorbed onto surrounding sediments (Pineau et al., 2008; Gabelle et al., 2012), which could act as a sink of aluminium to biota. Only a small part of Al in sediments is available (i.e., non-silicate-bound), involving reactions such as solubilization, polymerization, complexation, crystallization, which are strongly controlled by the pH conditions (Tam et al., 1989; Gunkel et al., 2011). Many studies have proved that Al toxicity is an important factor affecting the growth of plant and aquatic biota in acidic ecosystems (Meiri et al., 1993; Platt et al., 2001; Lu et al., 2011). However, detecting anthropogenic aluminium contamination could be especially difficult. Indeed, the usual trace metals contaminations studies (using background references comparisons) could not be easily transposed into a major element study (such as aluminum), because anthropological quantities are often negligible with regard to the natural amount. Partial extractions seem to provide more sensitive indication of aluminium contamination, as the first tests on natural sediments (Gabelle et al., 2012; Leleyter et al., 2012) highlighted some differences in the mineralogical partitioning of natural or anthropological aluminium.

The processes controlling the fate of dissolved aluminium are still not well understood (Wang et al., 2015) and up to now, the complex biogeochemical cycle of this element has been poorly understood (Gunkel et al., 2011). Benthic diatoms are an important component of phytobenthos inhabiting the intertidal mudflats by constituting the first primary producer in muddy estuaries (Underwood and Kronkamp, 1999) by biofilms at the sediment surface (Decho, 1990; Wingender et al., 1999). Benthic diatoms can be used as bioindicators of metallic pollution (Gold et al., 2003a; Lai et al., 2003; Cunningham et al., 2005; Morin et al., 2007), because of their high level of primary productivity in coastal sediments (Underwood and Kronkamp, 1999), their sensitivity to changes in water quality (Dixit et al., 1992; Stevenson and Pan, 1999), and their fundamental role in the food webs (Lefebvre et al., 2009). *In situ* studies conducted at sites exhibiting a high level of metals and microcosm experiments have demonstrated a decrease in productivity, diversity and changes in species composition of diatom communities (Takamura et al., 1989; Hill et al., 1997; Sabater, 2000). As reported by Gold et al. (2003b) and Duong et al. (2010), metal contamination had a strong effect on the density of diatom communities, possibly corresponding to a reduction in the rate of cell division of diatom species as demonstrated by Rivkin (1979), cessation or interruption of cell division (Dickman, 1998), the development of teratogenic forms (Arini et al., 2013), and also physiological impairments regarding photosynthesis, mitochondrial metabolism (Arini et al., 2012) and gene expression (Kim Tiam et al., 2012). Thus changes in benthic diatoms have been revealed to be a good estimator to assess heavy metal contamination (such as Cd, Cu, Zn, …) in freshwater ecosystems (Say and Whitton, 1980; Foster, 1982; Medley and Clements, 1998; Sabater, 2000). However, benthic diatoms in rivers can handle high concentrations of metal pollution such as cadmium, and biomonitoring surveys show the bioremediation process exerted by benthic diatoms (Arini et al., 2012) while reverse potential of teratogenic forms can show cadmium decontamination after experimental contamination (Arini et al., 2013). The behaviour of Al at the water-sediment interface is important for controlling dissolved concentrations in bottom waters via resuspension processes. The major processes removing dissolved Al from the water column include active biological uptake by diatoms (Wang et al., 2015), as benthic diatoms are ecosystem engineers capable of bioremediation of metallic pollution (Arini et al., 2012).

A mesocosm experiment was performed, with natural defaunated sediments that are contaminated with aluminium salts. Benthic diatoms are cultivated on these contaminated sediments in a tidal Mesocosm (Ubertini et al., 2015), in order to evaluate the aluminium impact on the growth of benthic diatom. The sediments are also leached by HCl (1M) to estimate the available/mobile part of the aluminium in the sediment. The aim of the present work is to study the potential uptake by benthic diatoms of non-natural Al in marine sediments, levels of toxicity for benthic diatoms and the capacity of the aluminium geochemical characterization procedures (total and labile) to be used in the evaluation of aluminum environmental risk.

2. Material and Methods

2.1. Sediments

A stock of fine sediment was collected in the Orne estuary (WGS84, 49.28°N, -0.24°W) from 10 to 20 cm below the surface and brought back to the lab. After 1 month in darkness, this sediment was sieved on a <1 mm mesh to remove the macro-fauna, then crushed and homogenized in an agate mortar and sieved at 80 µm. Control experiment consisted of Al natural sediment (T). Some aliquots of T-sediment were mixed thoroughly with variable quantities of aluminium sulphate ($Al_2(SO_4)_3 18,H_2O$, from Riedel-Haën, puriss quality, >99%), in order to obtain five Al-contaminated sediments, designed hereafter as C1 to C5 with added content of Al from 1 to 1000 $mg.kg^{-1}$ (Table 1). Then, contaminated sediments (C1 to C5) were sprayed with ultra-pure water and manually mixed. After drying (40°C), the C-sediments were crushed and homogenized in an agate mortar.

Al total contents in all the sediments were obtained after solubilisation of the solid matrix by an alkaline fusion procedure (NF ISO 14869-2, AFNOR 2002): 0.2g of dried sediment was thoroughly mixed with 0.8g of lithium metaborate and 0.2g of lithium tetraborate in a Pt crucible. The mixture was heated at 1000°C for 45 min and then the fusion product was immediately put in 60 mL of 1 mol.L^{-1} HNO_3 solution until total dissolution of the occurred residue. The recovered solutions were made up to a volume of 100 mL with 1 mol.L^{-1} HNO_3 solution.

The aluminium concentrations were measured by Inductively Coupled Plasma- Atomic Emission Spectrometry (ICP-AES, Varian, Vista MPX). The analytical quality of the chemical data was controlled using the standard certified material HR-1 (Canada Center for Inland Waters National Laboratory for Environmental) and the standard marine sediment (PACS2) (Gabelle et al., 2012; Hamdoun et al., 2015a).

Single extraction with 1 mol. L^{-1} HCl was also applied to all the sediments (T, C1 to C5). 1g of dry sediment was mixed with 10 mL of 1 mol.L^{-1} HCl solution under agitation (200tr/min) at room temperature for 1h. The resulting mixture was centrifuged (3000 rpm; 2min), filtered at 0.45 μm (syringe filter, Millex - SLHN033N). The obtained leachates were analyzed by ICP-AES.

Table 1. Total Al contents measured in the sediments after Al contamination, values are expressed per kg of dry sediment; SD: Standard Deviation (3 replicates); Normalisation ratio with [Cx] and [T] = total concentration in sediment Cx and T respectively

	Added Al *mg.kg^{-1}*	Measured total content *g.kg^{-1}*	*SD*	Normalisation ratio: [Cx]/[T]
T	0	21.7	*0.9*	1.00
C1	1	22.2	*0.1*	1.02
C2	10	22.2	*0.1*	1.02
C3	100	21.9	*0.1*	1.01
C4	500	22.7	*0.3*	1.05
C5	1000	22.8	*0.2*	1.05

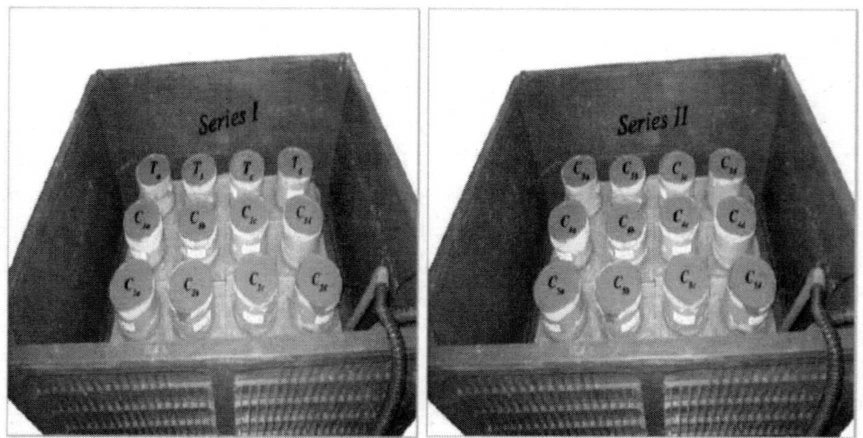

Figure 1. Unit (series I and II) with the 12 cores each.

2.2. Sediment Preparation and Benthic Diatom Cultivation Experiment in Response to Contamination by Aluminium

24 PVC cores (diameter: 15 cm, height: 20 cm) were filled with 2 layers: bottom layer with natural sediment and a superficial mud layer (1st cm of each core) that was treated with different levels of contamination (T and C1 to C5 sediments; 4 replicates) before being inoculated with fresh benthic diatoms. The first upper cm was then homogenized with an epipelic MPB inoculum collected from a mudflat located in the Orne estuary (WGS84, 49.28°N, -0.24°W) in Basse-Normandie during April 2011. It was collected by scratching the sediment surface. The biofilm was mainly composed of pennate diatoms including small *Navicula sp.* (length ~17μm, >95% of total MPB), *Amphora sp.*, *Pleurosigma sp.*, *Niztschia sp.* and *Cylindrotheca closterium*. The cores were positioned in trays (1.20 m x 1 m x1m, 12 cores capacity) (Figure 1). The first tray (called series I) contained T, C1 and C2 sediments, whereas the other tray (called series II) was performed with C3 to C5 sediments cores. The surface of the sediment was gently smoothed and cores were cultivated in a tidal mesocosm able to simulate a high/low tide alternation every 6 hours in order to simulate immersion and emersion phases.

The experimental mesocosm consists of two trays placed one on the top of the other (Figure 2). The lower unit served as seawater reservoir, to produce high and low tide, thanks to submersible pumps. The upper unit is used for the implementation of sediment cores (series I or II) and is covered with a roof equipped with a light intensity of 1600 µmol photons $m^2.s^{-1}$ (LUMINUX, 36W Osram), to reproduce day (during low tide) and night cycles. The experimental design was carried out according to Orvain et al. (2003) with 6h light: 18h dark regime. Natural seawater filtered over filters (1µM) was used for this experiment.

Water samples were collected at 0, 5 and 9 days and filtered through a 0.45 µm (Millipore) in polyethylene flask. The filtrates (14 mL) obtained were then acidified with HNO_3 (suprapur: 60%) and stored at 4°C for Al and trace metal analysis by Inductively Coupled Plasma Atomic Emission Spectrometry (ICP-AES, Varian, Vista MPX).

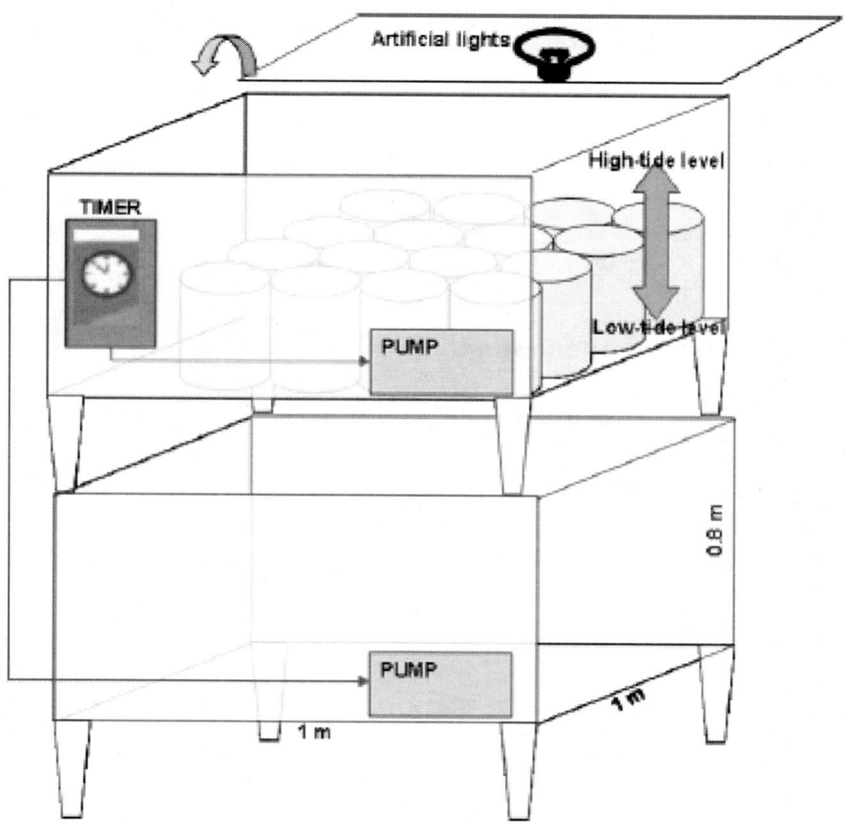

Figure 2. Tidal mesocosm reproducing the alternation of diurnal daytime low-tide period lasting 6 hours (8 :00AM – 2 : 00 PM), and nocturnal high-tide period lasting 18 hours (2:00 pm – 8:00).

Phytobenthic biomass was assessed by measuring the chlorophyll *a* (Chl *a*) content. Sediments samples were taken from the experimental cores at 5 and 9 days at the beginning of diurnal emersion periods in order to access respectively the growth and stationary phases of the biofilm (Orvain et al., 2003). The first upper cm of the sediment (2 replicates) was sampled with a 2-cm diameter cut syringe. The uppermost first centimeter was cut and mixed in 50 mL Falcon tubes. All frozen samples were lyophylised. Chl *a* extraction was applied on a given amount of dry sediment, with approximately 150 mg of sediment that was accurately weighed. Phytobenthic biomass was estimated by quantifying fluorometrically the chloropigments after extraction in 90% acetone and in the dark overnight under constant agitation at 4°C and centrifugation (4°C, 5 minutes, 3500 rpm). The chlorophyll extracts were measured on a Turner Designs TD 700 fluorometer (USA) following the method of Welschmeyer (1994). In order to avoid the dewatering over the emersion period (Perkins, 2003), water content and the bulk density of sediments could be used to express the Chl*a* as a content per m^{-2}.

3. Results and Discussion

3.1. Sediments Contamination

The measured quantities (Table 1) for all the contaminated sediments indicated that the contamination procedure was correct, without any loss or excessive contaminations. We noticed no meaningful difference (<5%, Table 1) between the Al total content in the background (T sediment) and the Al total content in contaminated sediments (low contamination with regard to the Al initial content which is naturally a major element in sediments).

Mobile metals concentrations are measured (3 replicates) in the tested sediments (T, C1 to C5, see Table 2) in the leachates after the HCl extraction and the mobile part is expressed as the percentage of the total Al content initial measured in the corresponding sediment. The quantities of labile Al remain quite stable (521 to 564 $mg.kg^{-1}$) for the low contamination rates (0 to 100 $mg.kg^{-1}$). On the opposite, the extracted quantities increased for the highest contamination rates (500 and 1000 $mg.kg^{-1}$), respectively 873 and 1311 $mg.kg^{-1}$, suggesting that an important part of the added aluminium is scavenged in the labile fraction of the sediment.

The resulting percentage of aluminium extracted by HCl 1M in contaminated sediments (C4 and C4, i.e., 500 or 1000 $mg.kg^{-1}$ of sediment) increased (respectively 4 and 6%) and thus doubled the quite stable percentage (2 to 3%) in sediments with low contamination rate (T and C1 to C3, from 0 to 100 mg.kg^{-1} of added Al).

Various authors (Agamian and Chau, 1976; Sutherland et al., 2001; Matus, 2007) reported that dilute HCl mainly extracted (from soils and sediments) Al species connected with anthropogenic contamination of the environment. Gabelle et al., (2012) also discriminated marine sediments contaminated by Al sacrificial from uncontaminated sediments thanks to similar results. They determined that Al mobilised by 1M HCl represented 2 ± 1% of the total Al present in the uncontaminated, whereas 5% to 11 ± 1% of the total Al was extracted by HCl from the sediments at the vicinity of Al sacrificial anodes.

Table 2. Quantities (Q_{HCl}) and percentages ($\%_{HCl}$) of Al extracted by HCl extraction, SD: standard deviation (3 replicates)

	(Q_{HCl}) mg.kg^{-1}	SD	($\%_{HCl}$)
T	564	*3*	3
C1	536	*2*	2
C2	521	*6*	2
C3	547	*19*	3
C4	873	*9*	4
C5	1311	*6*	6

3.2. Water Contamination

The average concentrations of the total Al determined in water throughout the 9 days of experimentation range from 0.04 to 0.07 mg.L^{-1} (Figure 3); these values are similar to concentrations previously reported for estuarine and marine waters. For example, concentrations from 9 to 70 μg/L, 5 to 50 μg/L, 10 to 60 μg/L, and 1 to 8 μg/L are reported for the GPMH (Seaport of Le Havre, France, Gabelle et al., 2012), Les Flamands (Seaport of Cherbourg, France, Mao et al., 2011) some Chinese estuaries (Zhang et al., 1999) and English Channel (Chou and Wollast, 1993) respectively.

The initial Al concentrations are similar for the two sets of experiment series (I and II). After 5 days, the concentrations increase in both cases, followed by a decrease for the next 4 days. The decrease is more pronounced for the low contamination set (0 to 10 mg.kg^{-1} of added Al) with Al concentrations in water lower than the initial ones, whereas the final concentrations remain higher than the initial ones for the second set (100 to 1000 mg.kg^{-1} of added Al).

The increase noticed after 5 days is probably due to the partial solubilisation of the added aluminum salts. The decrease then observed after 9 days could result from various processes, such as re-adsorption by the sediments, or uptake by the diatoms. Indeed, diatom can use a significant quantity of aluminium for their physiological requirements, but phytobenthic diatoms are above all capable of bioremediating pollutants out of their cells, by producing anionic polymers in the microenvironment (Wingender et al., 1999). These Extracellular Polymeric Substances (EPS) are very efficient to sequester cationic pollutants such as heavy metals, due to the anionic nature of these compounds (Decho, 2000). In case of physiological and environmental stress (contamination by heavy metals, pesticides, sediment evaporation, dehydration, saline stress, nutrient absence), a wide quantity of the photoassimilates of the benthic diatoms are secreted at the water/sediment interface (Orvain et al., 2003; McKew et al., 2011) to preserve cells against pollutants.

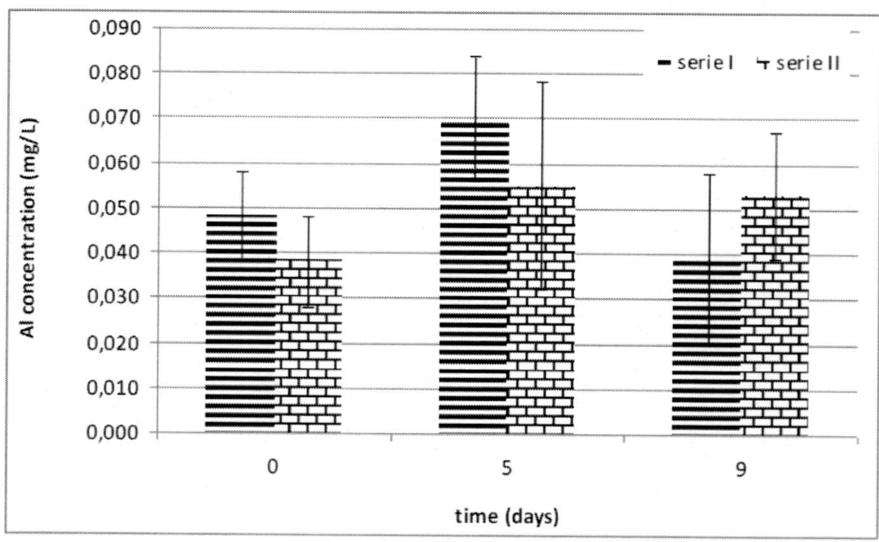

Figure 3. Al concentrations in the water throughout the experiment, 5 replicates.

3.3. Impact of Aluminium Contamination on Benthic Diatom Growth

The chl*a* content measured on the control cores (non contaminated T-sediment) remains stable from 5 to 9 days at 14 $\mu g.g^{-1}$ (dry weight sediment, Figure 4) (or 120 $mg.m^{-2}$) because benthic diatoms show a rapid growth phase in 5 days (or less). The Chl*a* measured after 5 days are similar to the control for the two lowest contaminated samples (C1 and C2, with 1 and 10 $mg.kg^{-1}$ of added Al). However, after 9 days, the values for C1 and C2 differ from the control: the higher the Al content is, the higher the Chl*a* is. This reflects the stimulation of the growth of benthic diatoms when there is moderately contamination in aluminium. Chl*a* content increases between T and C2 after 9 days of culture.

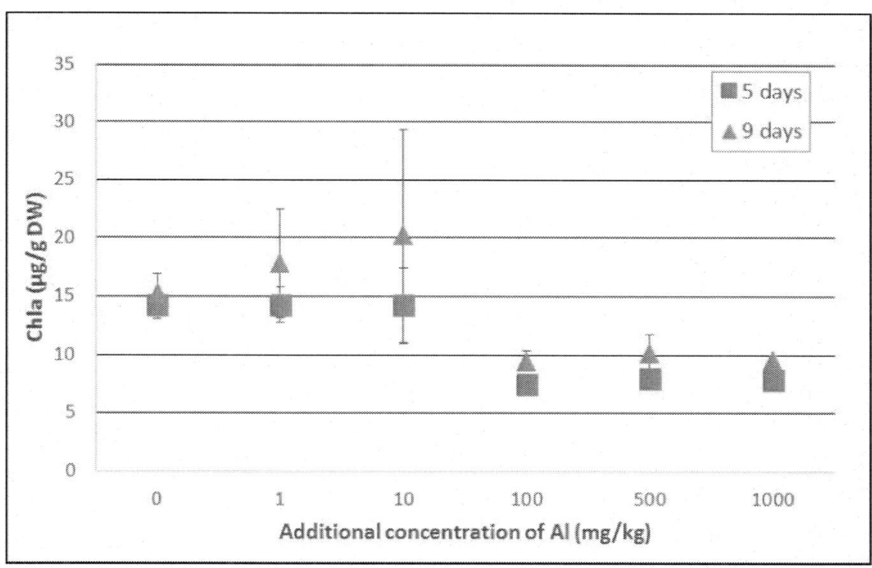

Figure 4. Influence of increasing stress of additional aluminium on Chlorophyll *a* (Chl *a*) contents in sediments.

In case of a moderate Al contamination (≤10 $mg.kg^{-1}$), benthic diatoms should resist to the presence of heavy metal under an apparent threshold of 10 $mg.kg^{-1}$. In case of physiological stress and pollution, benthic diatoms are known to secrete EPS in their micro-environment; the stimulation of this process can eventually lead to a better growth of the biofilm. We can hypothesize that benthic diatoms must enhance their photosynthesis activity, productivity and EPS secretion, to exert a phytoremediation process, that is mandatory to resist to the heavy metal contamination. Thus a moderate contamination can have a positive effect on the net growth of benthic diatoms in the epipelic biofilm: these preliminary results must be further investigated to refine the bioremediating strategy of these microalgae In contrast to the algal growth at low contamination rate (T, C1 and C2, ie, 0 to10 $mg.kg^{-1}$ of added Al), our results indicate a lack of development of microalgae at higher contamination level ($\geq$ 100 $mg.kg^{-1}$ of added Al, sediments designed as C3, C4 and C5). In agreement with our results, various authors reported the same observations on the inhibitory effects of metals on algal growth capacity (Prasad and Prasad, 1982; Lasheen et al., 1989; Payne and Price, 1999; Nayar et al., 2003; Duong et al., 2010). Benthic diatoms do not grow and Chl*a* concentration remains very low, without any difference between days 5 and 9, revealing that no resilience can occur, even after some days after a strong contamination by heavy metals. The negative effects are so drastic, that this suggests a mass mortality of benthic diatoms, in response to high levels of Al contamination over 100 $mg.kg^{-1}$. Resilience strategies in response to Al contamination are efficient up to the threshold value of 10 $mg.kg^{-1}$, and this process surprisingly stimulates the growth and probably also the physiological activities in the biofilm.

CONCLUSION

This experiment has proved that the Al-contamination of sediments has a strong effect on diatom communities growing at the muddy sediments surface. The negative effect can be detected beyond 10 mg of aluminium salts added by kg of natural sediment. Indeed, benthic diatoms show a lethal threshold of aluminium contamination of 10 $mg.kg^{-1}$ of sediment. However, at lowest concentrations of added aluminium, phytobenthic are capable of resilience and a stimulation of the growth is observed probably in relation to a stimulation of photosynthetic activity, to produce and excrete more EPS in the biofilm that can absorb aluminium in excess. Further studies must be performed to better refine physiological mechanisms of benthic diatoms to exert a bioremediation of contaminating heavy metals and especially aluminium.

HCl-extraction showed that at high level of Al-salts contamination (500 mg or 1000 mg per kg of sediment), the aluminum mobility increases, respectively of 4 and 6%, compared to the initial values in natural conditions

(3%). The Al mobility remained unchanged for the lowest contamination rates (1, 10 or 100 mg of salt per kg of sediment). The 1M HCl-extraction, widely used in the literature for metals mobility evaluation in marine sediments, underestimates aluminium availability for phytobenthic communities, in such case. Indeed, the% HCl for the C2 sediment (100 $mg.kg^{-1}$ of added Al), is similar to the ones obtained for lower contamination levels, whereas drastic effects on the diatoms are only observed for this specific concentration rate. On the opposite the increase of the% HCl for the highest contamination level (100 $mg.kg^{-1}$ of added Al) can be considered as a good prediction of the negative impact of Al on the diatoms that is actually observed.

Finally, in case of aluminium contamination, HCl extraction can be helpful for risk assessment only for significant contamination levels, that induces HCl mobility superior to 4%. When lower% HCl is measured complementary investigations might be necessary. Thus, an abnormal result (over 3%) for Al lability (after 1M HCl extraction) allows classifying the studied sediment as Al contaminated with important negative effects for benthic diatom. However a more 'classic' result (≤3%), does not allow the discrimination of contaminated or uncontaminated sediments.

REFERENCES

Allègre C-J. and Michard G., 1973, *Introduction à la géochimie*, eds: Presses Universitaires de France, 220 p.

Agemian H. and Chau A. S. Y., 1976, Evaluation of extraction techniques for the determination of metals in aquatic sediments. *The Analyst,* 101; 761-767.

Alvarenga P., Simões I., Palma P., Amaral O., Matos J. X., 2014, Field study on the accumulation of trace elements by vegetables produced in the vicinity of abandoned pyrite mines. *Science of the Total Environment,* 470–471; 1233–1242.

Arini A., Feurtet-Mazel A., Morin S., Maury-Brachet R., Coste M., Delmas F., 2012, Remediation of a watershed contaminated by heavy metals: a 2-year field biomonitoring of periphytic biofilms. *Science of the Total Environment,* 425; 242-253.

Arini A., F. Durant F., Coste M., Delmas F., Feurtet-Mazel A., 2013, Cadmium decontamination and reversal potential of teratological forms of the diatom *Planothidium frequentissimum* (Bacillariophyceae) after experimental contamination. *Journal of Phycology,* 49 (2); 361-370.

Burton E. D., Phillips I. R., Hawker D. W., 2005, Reactive sulfide relationships with trace metal extractability in sediments from southern Moreton Bay Australia. *Marine Pollution Bulletin, 50*; 583–608.

Chand V., Prasad S., 2013, ICP-OES assessment of heavy metal contamination in tropical marine sediments: A comparative study of two digestion techniques. *Microchemical Journal*, 111; 53-61.

Chou L. and Wollast R., 1993, Distribution and origin of dissolved aluminium in the English Channel. Oceanologica Acta, 16; 577-583.

Covelli S., Fantolan G., 1997, Application of normalization procedure in determining regional geochemical baseline. *Environmental Geology,* 30; 34–45.

Cunningham L., Raymond B., Snape L., Riddle M.J., 2005, Benthic diatom communities as indicators of anthropogenic metal contamination at Casey Station, Antarctica. *Journal of Paleolimnology,* 33; 499–513.

Decho A. W., 1990, Microbial exopolymer secretions in ocean environments: their role(s) in food webs and marine processes. *Oceanography Marine Biology: An Annual Review,* 28; 73–153.

Decho AW., 2000, Microbial biofilms in intertidal systems: an overview. *Continental and Shelf Research,* 20; 1257-1273.

Devesa-Rey R., Díaz-Fierros F., Barral M.T., 2010, Trace metals in river bed sediments: An assessment of their partitioning and bioavailability by using multivariate exploratory analysis. *Journal of Environmental Management,* 91; 2471-2477.

Dickman MD., 1998, Benthic marine diatoms deformities associated with contaminated sediments in Hong Kong. *Environment International,* 24; 749–759.

Dixit SS., Smol JP., Kingston JC., Charles DF., 1992, Diatoms: powerful indicators of environmental change. *Environmental Science and Technology,* 26; 23–33.

Doherty G. B., Brunskill G. J., Ridd M. J., 2000, Natural and enhanced concentrations of trace metals in sediments of cleveland bay, Great Barrier Reef Lagoon, Australia. *Marine Pollution Bulletin,* 41; 337–344.

Dubrulle C., Lesueur P., Boust D., Dugué O., Poupinet N., Lafite R., 2007, Source discrimination of fine-grained deposits occurring on marine beaches: the Calvados beaches (eastern Bay of the Seine, France), Estuarine. *Coastal Shelf Sciences,* 72; 138–154.

Duong T- T., Morin S., Coste M., Herlory O., Feurtet-Mazel A., Boudou A., 2010, Experimental toxicity and bioaccumulation of cadmium in freshwater periphytic diatoms in relation with biofilm maturity. *Science of the Total Environment,* 408; 552–562.

El Azzi D., Viers J., Guiresse M., Probst A., Aubert D., Caparros J., Charles F., Guizien K., Probst J.L., 2013, Origin and fate of copper in a small Mediterranean vineyard catchment: New insights from combined chemical extraction and δ^{65}Cu isotopic composition. *Science of the Total Environment,* 463–464; 91–101.

Foster PL., 1982, Species associations and metal contents of algae from river polluted by heavy metals. *Freshwater Biology,* 2; 17–39.

Gabelle C., Baraud F., Biree L., Gouali S., Hamdoun H., Rousseau C., Van Veen E. and Leleyter L, 2012, The impact of aluminium sacrificial anodes on the marine environment: a case study. *Applied Geochemistry,* 27; 2088-2095.

Giancoli Barreto S. R., Nozaki J., De Oliveira E., Do Nascimento Filho V. F., Aragão P. H. A., Scarminio I. S., Barreto W. J., 2004, Comparison of metal analysis in sediments using EDXRF and ICP-OES with the HCl and Tessier extraction methods. *Talanta,* 64; 345–354.

Gold C., Feurtet-Mazel A., Coste M., Boudou A., 2003a, Impacts of Cd and Zn on the development of periphytic diatom communities in artificial streams located along a river pollution gradient. *Archives of Environmental Contamination and Toxicology,* 44; 189–97.

Gold C., Feurtet-Mazel A., Coste M., Boudou A., 2003b. Effects of cadmium stress on periphytic diatom communities in indoor artificial streams. *Freshwater Biology,* 148; 316–28.

Gunkel G., Beulker C;, Gernet U., Viteri F. 2011, Aluminium in lake Cuicocha, Ecuador, an Andean Crater Lake: filterable, gelatinous and microcrystal Al occurrence. *Aquatic Geochemistry,* 17; 109-127.

Hamdoun H., Van-Veen E., Basset B., Lemoine M., Coggan J., Leleyter L., Baraud F., 2015a, Characterization of harbor sediments from the English Channel: assessment of heavy metal enrichment, biological effect and mobility. *Marine Pollution Bulletin, 90*; 273-280.

Hamdoun H., Leleyter L., Van-Veen E., Basset B., Lemoine M., Baraud F., 2015b, Comparison of three procedures (single, sequential and kinetic extraction) for mobility assessment of Cu, Pb and Zn from harbour sediments. *CRAS-Géosciences,* doi:10.1016/j.crte.2015.03.003, in press.

Hill BH., Lazorchak JM., McCormick FH., Willingham WT., 1997, The effects of elevated metals on benthic community metabolism in a rocky mountain stream. *Environmental Pollution,* 96; 183–90.

Islam MS., Ahmed MK., Habibullah-Al-Mamun M., 2015, Metal speciation in soil and health risk due to vegetables consumption in Bangladesh. *Environmental Monitoring and Assessment,* 187; 1-15.

Kim Tiam S., Feurtet-Mazel A., Delmas F., Mazzella N., Morin S., Daffe G., Gonzalez P., 2012, Development of q-PCR approaches to assess water quality: Effects of cadmium on gene expression of the diatom Eolimna minima. *Water Research,* 46; 934-942.

Lai SD., Chen PC., Hsu HK., 2003. Benthic algae as monitor of heavy metals in various polluted rivers by energy dispersive X-ray spectrometer. *Journal of Environmental Science and Health,* 38; 855–66.

Larner B. L., Palmer A. S., Seen A. J., Townsend A. T., 2008, A comparison of an optimised sequential extraction procedure and dilute acid leaching of elements in anoxic sediments including the effects of oxidation on sediment metal partitioning. *Analytica Chimica Acta,* 608; 147–157.

Lasheen M., Shehata S., Hali G., 1989, Effects of cadmium, copper and chromium (VI) on the growth of Nile water algae. *Water, Air, and Soil Pollution,* 50; 19–30.

Lefebvre S., Marín Leal JC., Dubois S., Orvain F., Blin J-L., Bataillé M-P., Ourry A., Galois R, 2009, Seasonal dynamics of trophic relationships among co-occurring suspension-feeders in two shellfish culture dominated ecosystems. *Estuarine Coastal and Shelf Science,* 82; 415-425.

Leleyter L., Rousseau C., Biree L. and Baraud F., 2012, Comparison of EDTA, HCl and sequential extraction procedures, for selected metals (Cu, Mn, Pb, Zn), in soils, riverine and marine sediments. *Journal of Geochemical Exploration,* 116–117; 51–59.

Leleyter L. and Baraud F., 2005, Evaluation of metals mobility in sediments by single or sequential extractions. *Comptes Rendus Geosciences,* 337; 571-579.

Leleyter L. and Probst J. L., 1999, A new sequential extraction procedure for the speciation of particulate trace elements in river sediments. *International Journal of Environmental Analytical Chemistry,* 73 (2); 109-128.

Lu W., Ma Y., Lin C. 2011, Status of aluminium in environmental compartments contaminated by acidic mine water. *Journal of Hazardous Waste Materials,* 189; 700-709.

Mao A., Mahaut M-L., Pineau S., Barillier D., Caplat C., 2011, Assessment of sacrificial anode impact by aluminum accumulation in mussel Mytilus edulis: A large-scale laboratory test. *Marine Pollution Bulletin,* 62; 2707–2713.

Matus P., 2007, Evaluation of separation and determination of phytoavailable and phytotoxic aluminium species fractions in soils, sediment and water samples by five different methods. *Journal of inorganic biochemistry,* 101; 1214-1223.

Meiri H., Banin E., Roll M., Rouesseau A., 1993, Toxic effects of aluminium on nerve cell and synaptic transmission. *Progress in Neurobiology,* 89; 121.

McKew B. A., Taylor J. D., McGenity T. J., Underwood, G. J., 2011, Resistance and resilience of benthic biofilm communities from a temperate saltmarsh to desiccation and rewetting. *The ISME Journal,* 5; 1–30.

Medley NC., Clements WH., 1998, Responses of diatom communities to heavy metals in Rocky Mountain streams: the influence of longitudinal variation. *Ecological Applications,* 8; 631–44.

Meybeck M., Lestel L., Bonté P., Moilleron R., Colin J.-L., Rousselot O., Hervé D., De Pontevès C., Grosbois C., Thévenot, D. R., 2007, Historical perspective of heavy metals contamination (Cd, Cr, Cu, Hg, Pb, Zn) in the Seine River basin (France) following a DPSIR approach (1950–2005). *Science of the Total Environment,* 375; 204–231.

Morin S., Vivas-Nogues M., Duong TT., Boudou A., Coste M., Delmas F., 2007, Dynamics of benthic diatom colonization in a cadmium/zinc-polluted river (Riou-Mort, France). *Fundamental and Applied Limnology,* 168; 179–87.

Nayar S., Goh BPL., Chou LM., Reddy S., 2003, In situ microcosms to study the impact of heavy metals resuspended by dredging on periphyton in a tropical estuary. *Aquatic Toxicology,* 64; 293–306.

Orvain F., Galois R., Barnard C., Sylvestre A., Blanchard G., Sauriau PG., 2003, Carbohydrate production in relation to microphytobenthic biofilm development: An integrated approach in a tidal mesocosm. *Microbial Ecology,* 45; 237-251.

Payne CD., Price NM., 1999, Effects of cadmium toxicity on growth and elemental composition of marine phytoplankton. *Journal of Physiology,* 35; 239–302.

Peña-Icart M., Mendiguchía C., Villanueva-Tagle M E., Pomares-Alfonso M S., Moreno C., 2014, Revisiting methods for the determination of bioavailable metals in coastal sediments. *Marine Pollution Bulletin,* 89; 67-74.

Perkins R., 2003, Changes in microphytobenthic chlorophyll a and EPS resulting from sediment compaction due to de-watering: opposing patterns in concentration and content. *Continental Shelf Research,* 23; 575–586.

Pineau S., Gabelle C., Mao A., Baraud F., Caplat C., Leleyter L., Masson D., 2008, Report presenting environmental issues- analytical results and ecotoxicological impact – related to infrastructure protection, rapport de recherche, Efforts – WP 2.2; p 87.

Platt B., Fiddler G., Riedel G., Henderson Z., 2001, Aluminium toxicity in the rat brain: histochemical and immunocytochemical evidence. *Brain Research Bulletin,* 55; 257–267.

Prasad PV., Prasad PS., 1982, Effects of cadmium, lead and nickel on three freshwater green algae. *Water Air and Soil Pollution,* 17; 263–8.

Rivkin RB., 1979, Effects of lead on growth of the marine diatom Skeletonema costatum. *Marine Biology,* 50; 239–47.

Rousseau C., Baraud F., Leleyter L., Gil O. 2009, Cathodic protection by zinc sacrificial anodes: Impact on marine sediment metallic contamination. *Journal of Hazardous Materials,* 167; 953-958.

Roussiez V., Probst A., Probst J-L., 2013, Significance of floods in metal dynamics and export in a small agricultural catchment. *Journal of Hydrology,* 499; 71–81.

Sabater S., 2000, Diatom communities as indicators of environmental stress in the Guadiamar River, S–W Spain, following a major mine tailing spill. *Journal of Applied Physiology,* 12; 113–24.

Say PJ., Whitton BA., 1980, Changes in flora downa streamshowing a zinc gradient. *Hydrobiologia,* 76; 255–62.

Scouller R. C., Snape I., Stark J., Gore D. B., 2006, Evaluation of geochemical methods for discrimination of metal contamination in Antartic marine sediments: a case study from Casey Station. *Chemosphere,* 65; 294–309.

Snape I., Scouller R. C., Stark S. C., Stark J., Riddle M. J., Gore D. B., 2004, Characterisation of the dilute HCl extraction method for the identification of metal contamination in Antarctic marine sediments. *Chemosphere,* 57; 491–504.

Stevenson RJ., Pan Y., 1999, Assessing ecological conditions in rivers and streams with diatoms. In: Stoermer EF, Smol JP, editors. *The diatoms: applications to the environmental and earth sciences.* Cambridge, UK: Cambridge University Press, 11–40.

Sutherland R. A., Tack F. M. G., Tolosa C. A., Verloo M. G., 2001, Metal Extraction from Road Sediment using Different Strength Reagents: Impact on Anthropogenic Contaminant Signals. *Environmental Monitoring and Assessment,* 71; 221-242.

Sutherland R. A., 2002, Comparison between non-residual Al, Co, Cu, Fe, Mn, Ni, Pb and Zn released by a three-step sequential extraction procedure and a dilute hydrochloric acid leach for soil and road deposited sediment. *Applied Geochemistry,* 17; 353-365.

Takamura N., Kasai F., Watanabe MM., 1989, Effects of Cu, Cd and Zn on photosynthesis of freshwater benthic algae. *Journal of Applied Physiology,* 1; 39–52.

Tam NFY., Wong Y. S., Wong M. H., 1989, Effects of acidity on acute toxicity of aluminium-waste and aluminium contaminated soil. *Hydrobiologia,* 188/189; 385-395.

Tessier A., Campbell P. G. C., Bisson M., 1979, Sequential extraction procedure for speciation of particulate trace metals. *Analytical Chemistry,* 51; 844–851.

Ubertini M., Lefebvre S., Rakotomalala C., Orvain F., 2015, Impact of sediment grain-size and biofilm age on epipelic microphytobenthos resuspension. Journal *of Experimental Marine Biology and Ecology,* 467; 52-64.

Underwood G. J. C., Kronkamp J., 1999, Primary production by phytoplankton and microphytobenthos in estuaries. *Advances in Ecological Research,* 29; 93-153.

Wang Z-W., Ren J-L., Zhang G-L., Liu S-M., Zhang X-Z., Liu Z., Zhang J., 2015, Behavior of dissolved aluminium in the Huanfhe (Yellow River) and its estuary: Impact of human activities and sorption processes. *Estuarine, Coastal and Shelf Science,* 153; 86-95.

Welschmeyer NA., 1994, Fluorometric analysis of chlorophyll a in the presence of chlorophyll b and pheopigments. *Limnology and Oceanography,* 39; 1985−1992.

Wingender J., Neu T. R. and Flemming H. C., 1999, What are bacterial extracellular polymeric substances? In *'Microbial Extracellular Polymeric Substances: Characterization, Structure and Function'* (Eds J. Wingender, T. R. Neu, and H. C. Flemming.), 1- 15.

Zhang J., Xu H., Yu Z. G., Wu Y., Li J. F., 1999, Dissolved aluminium in four Chinese estuaries: evidence of biogeochemical uncoupling of Al with nutrients. *Journal of Asian Earth Sciences,* 17; 333–343.

In: Marine Sediments
Editor: Shirley Williams

ISBN: 978-1-63485-127-5

Chapter 4

BIBLIOGRAPHY

A biogeographic assessment of seabirds, deep sea corals and ocean habitats of the New York Bight: science to support offshore spatial planning LCCN: 2012452598 Main title: A biogeographic assessment of seabirds, deep sea corals and ocean habitats of the New York Bight: science to support offshore spatial planning/repared by Center for Coastal Monitoring and Assessment, Biogeography Branch; editors, Charles Menza, Brian P. Kinlan, Dan S. Dorfman, Matthew Poti, Chris Caldow. Published/Created: Silver Spring, Md.: U.S. Department of Commerce, National Oceanic and Atmospheric Administration, National Ocean Service, National Centers for Coastal Ocean Science, 2012. Description: ii, 224 p.: col. ill., col. maps; 28 cm. Links: Online version via the NOAA Central Library. http://docs.lib.noaa.gov/noaa_documents/NOS/NCCOS/NY-Bight-report/ http://docs.lib.noaa.gov/noaa_documents/NOS/NCCOS/NY-Bight-report/nyc msp_execsumm.pdf Executive summary in PDF LC classification: QH541.5.C65 B564 2012 Related names: Menza, Charles W. (Charles William), editor. Kinlan, Brian P. (Brian Patrick), editor. Dorfman, Dan S., editor. Poti, Matthew, editor. Caldow, Chris, editor. Center for Coastal Monitoring and Assessment (U.S.). Biogeography Program, creator. National Centers for Coastal Ocean Science (U.S.), sponsor, publisher. Subjects: Sea birds--New York Bight (N.J. and N.Y.) Sea birds--New York Bight (N.J. and N.Y.)--Geographical distribution. Sea birds--New York Bight (N.J. and N.Y.)--Geographical distribution--Maps. Deep sea corals--New York Bight (N.J. and N.Y.) Deep sea corals--Habitat--New York Bight (N.J. and N.Y.) Ocean temperature--New York Bight (N.J. and N.Y.) Marine sediments--New York Bight (N.J. and N.Y.) Notes: "April 2012." Format not distributed to depository libraries. Includes bibliographical references. Additional formats:

Online version: biogeographic assessment of seabirds, deep sea corals and ocean habitats of the New York Bight (OCoLC)844735185 Series: NOAA technical memorandum NOS NCCOS; 141 NOAA technical memorandum NOS NCCOS; 141.

An assessment of chemical contaminants, toxicity and benthic infauna in sediments from the St. Thomas East End Reserves (STEER) LCCN: 2014407153 Personal name: Pait, Anthony S. Main title: An assessment of chemical contaminants, toxicity and benthic infauna in sediments from the St. Thomas East End Reserves (STEER) / Anthony S. Pait, S. Ian Hartwell, Andrew L. Mason, Robert A. Warner, Christopher F.G. Jeffrey, Anne M. Hoffman, Dennis A. Apeti, Francis R. Galdo, Jr., and Simon J. Pittman. Published/Produced: [Silver Spring, Maryland]: NOAA NCCOS Center for Coastal Monitoring and Assessment, [2013] Description: vi, 70 pages; 28 cm. LC classification: QH545.C59 P35 2013 Subjects: Contaminated sediments--United States Virgin Islands--Saint Thomas--Analysis. Marine sediments--Toxicology--United States Virgin Islands--Saint Thomas--Analysis. Benthic animals--Effect of chemicals on--United States Virgin Islands--Saint Thomas. Toxicity testing--United States Virgin Islands--Saint Thomas. Notes: "May 2013." Includes bibliographical references (pages 47-50). Series: NOAA technical memorandum NOS NCCOS; 156

An introduction to hydraulics of fine sediment transport LCCN: 2013000653 Personal name: Mehta, Ashish J. Main title: An introduction to hydraulics of fine sediment transport / Ashish J. Mehta. Published/Produced: New Jersey: World Scientific, [2014] Description: xix, 1039 pages: illustrations; 24 cm. ISBN: 9789814449489 (hbk.: alkaline paper) LC classification: TC175.2 .M44 2014 Contents: Fine Sediment Classification and Characteristic Properties -- Processes of Flocculation, Settling, Deposition, Consolidation, Gelation and Erosion -- Properties and Behavior of Fluid Mud -- Wave-Mud Interaction -- Sedimentation Problems in Coastal and Estuarine Waters as well as Lakes, and Small Tidal Basins. Subjects: Sediment transport. Hydraulic engineering. Marine engineering. Coastal engineering. Coastal sediments. Estuarine sediments. Marine sediments. Notes: Includes bibliographical references (pages 975-1026) and index. Series: Advanced series on ocean engineering; volume 38

Anoxia: evidence for eukaryote survival and paleontological strategies LCCN: 2011935457 Main title: Anoxia: evidence for eukaryote survival and paleontological strategies / edited by Alexander V. Altenbach, Joan M. Bernhard, and Joseph Seckbach. Published/Created: Dordrecht: Springer, c2012. Description: xxxv, 648 p.: ill., ports. (some col.), maps (some col.); 24

cm. Links: Publisher description http://www.loc.gov/catdir/enhancements /fy1316/2011935457-d.html Table of contents only http://www.loc.gov /catdir/enhancements/fy1316/2011935457-t.html ISBN: 9789400718951 (alk. paper) 9400718950 (alk. paper) 9789400718968 (e-ISBN) 9400718969 (e-ISBN) LC classification: QH518.5 .A56 2011 Related names: Altenbach, Alexander V. Bernhard, Joan M. Seckbach, J. (Joseph) Contents: Anaerobic eukaryotes / T. Fenchel -- Biogeochemical reactions in marine sediments underlying anoxic water bodies/ T. Treude -- Diversity of anaerobic prokaryotes and eukaryotes: breaking long-established dogmas / A. Oren -- The biochemical adaptations of mitochondrion-related organelles of parasitic and free-living microbial eukaryotes to low oxygen environments / A.D. Tsaousis et al. -- Hydrogenosomes and mitosomes: mitochondrial adaptations to life in anaerobic environments, R.M. De Graaf and J.H.P. Hackstein -- Adapting to hypoxia: lessons from vascular endothelial growth factor / N.S. Levy and A.P. Levy -- Magnetotactic protists at the oxic-anoxic transition zones of coastal aquatic environments / D.A. Bazylinski et al. -- A novel ciliate (Ciliophora: Hypotrichida) isolated from bathyal anoxic sediments / D.J. Baudoin et al. -- The wood-eating termite hindgut: diverse cellular symbioses in a microoxic to anoxic environment / M.F. Dolan -- Ecological and experimental exposure of insects to anoxia reveals surprising tolerance / W.W. Hoback -- The unusual response of encysted embryos of the animal extremophile, Artemia franciscana, to prolonged anoxia / J.S. Clegg -- Survival of tardigrades in extreme environments: a model animal for astrobiology / D.D. Horikawa -- Long-term anoxia tolerance in flowering plants / R.M.M. Crawford -- Benthic Foraminifera: inhabitants of low-oxygen environments / K.A. Koho and E. Piña-Ochoa -- Ecological and biological response of benthic Foraminifera under oxygen-depleted conditions: evidence from laboratory approaches / P. Heinz and E. Geslin -- The response of benthic Foraminifera to low-oxygen conditions of the Peruvian oxygen minimum zone / J. Mallon et al. -- Benthic foraminiferal communities and microhabitat selection on the continental shelf off central Peru / J. Cardich et al. -- Living assemblages from the "dead zone" and naturally occurring hypoxic zones / K.R. Buck et al. -- The return of shallow shelf seas as extreme environments: anoxia and macrofauna reactions in the northern Adriatic Sea / M. Stachowitsch et al. -- Meiobenthos of the oxic/anoxic interface in the south-western region of the Black Sea: abundance and taxonomic composition / N.G. Sergeeva et al. -- The role of eukaryotes in the anaerobic food web of stratified lakes / A. Saccà -- The anoxic Framvaren Fjord as a model system to study protistan diversity and evolution / T. Stoeck and A. Behnke --

Characterizing an anoxic habitat: sulfur bacteria in a meromictic alpine lake / G.B. Fritz et al. -- Ophel, the newly discovered hypoxic chemolitho-autotrophic groundwater biome: a window to ancient animal life / F.D. Por -- Microbial eukaryotes in the marine subsurface? / V.P. Edgcomb and J.F. Biddle -- On the use of stable nitrogen isotopes in present and past anoxic environments / U. Struck -- Carbon and nitrogen isotopic fractionation in Foraminifera: possible signatures from anoxia / A.V. Altenbach et al. -- The functionality of pores in benthic Foraminifera in view of bottom water oxygenation: a review / N. Glock et al. -- Anoxia-dysoxia at the sediment-water interface of the southern Tethys in the late Cretaceous: Mishash Formation, southern Israel / A. Almogi-Labin et al. -- Styles of agglutination in benthic Foraminifera from modern Santa Barbara Basin sediments and the implications of finding fossil analogs in Devonian and Mississippian black shales / J. Schieber -- Did redox conditions trigger test templates in Proterozoic Foraminifera? / A.V. Altenbach and M. Gaulke -- The relevance of anoxic and agglutinated benthic Foraminifera to the possible Archean evolution of eukaryotes / W. Altermann et al. Subjects: Anoxic zones. Eukaryotic cells--Evolution. Adaptation (Physiology) Extreme environments--Microbiology. Anaerobiosis. Micropaleontology. Anaerobiosis. Eukaryota. Adaptation, Physiological. Biological Evolution. Paleontology. Notes: Includes bibliographical references and indexes. Series: Cellular origin, life in extreme habitats and astrobiology, 1566-0400; v. 21 Cellular origin and life in extreme habitats and astrobiology; v. 21. 1566-0400

Assessment of heavy metal contamination in the marine environment of the Arabian Gulf LCCN: 2012048209 Personal name: Naser, Humood Abdulla. Main title: Assessment of heavy metal contamination in the marine environment of the Arabian Gulf / Humood Abdulla Naser. Published/Produced: Hauppauge, New York: Novinka, [2013] Description: viii, 99 pages: color illustrations, maps; 23 cm. ISBN: 9781624176197 (soft cover) LC classification: GC1451 .N36 2013 Contents: General Introduction -- Valued Ecosystem Components in the Arabian Gulf -- Anthropogenic Sources of Heavy Metals in the Arabian Gulf -- Heavy Metal Levels in Algal Species from the Arabian Gulf -- Heavy Metal Levels in Molluscs from the Arabian Gulf -- Heavy Metal Levels in Fish Species from the Arabian Gulf -- Heavy Metal Concentrations in Seawaters from the Arabian Gulf -- Heavy Metal Concentrations in Sediments from the Arabian Gulf -- Macrobenthic Community Structure as a Bioindicator for Heavy Metal Contamination -- Spatial Distribution of Heavy Metals in Marine Sediments Influenced by Land-based Anthropogenic Sources in Bahrain: A Case Study -- Management of Heavy Metals in the Arabian Gulf -- Conclusions and Recommendations. Subjects: Marine pollution--Persian Gulf. Heavy metals--Environmental aspects--Persian Gulf. Persian Gulf--Environmental conditions. Notes: Includes bibliographical references and index. Series: Environmental Health-Physical, Chemical and Biology Factors

Carbonates: sedimentology, geographical distribution and economic importance LCCN: 2013036362 Main title: Carbonates: sedimentology, geographical distribution and economic importance / Bailey A. Hughes and Thompson C. Wagner, editors. Published/Produced: New York: Novinka, [2013] ©2013 Description: ix, 93 pages: illustrations, map; 23 cm. ISBN: 9781629481784 (soft cover) LC classification: QE471.15.C3 C393 2013 Related names: Hughes, Bailey A., editor of compilation. Wagner, Thompson C., editor of compilation. Subjects: Carbonates. Marine sediments. Notes: Includes bibliographical references and index. Series: Chemical engineering methods and technology Geology and mineralogy research developments

Deep-marine systems: processes, deposits, environments, tectonics and sedimentation LCCN: 2015007967 Personal name: Pickering, K. T. (Kevin T.) Main title: Deep-marine systems: processes, deposits, environments, tectonics and sedimentation / by Kevin T. Pickering & Richard N. Hiscott; with contribution from Thomas Heard. Published/Produced: Chichester, West Sussex; Hoboken, NJ: John Wiley & Sons Inc., [2015] Description: pages cm ISBN: 9781118865491 (cloth) 9781405125789 (pbk.) LC classification: GC380.15 .P54 2015 Related names: Hiscott, Richard N. Subjects: Marine sediments. Plate tectonics. Notes: Includes bibliographical references and index. Additional formats: Online version: Pickering, K. T. (Kevin T.) Deep-marine systems Chichester, West Sussex; Hoboken, NJ: John Wiley & Sons Inc., [2015] 9781118865422 (DLC) 2015017033

Marine geochemistry LCCN: 2012010712 Personal name: Chester, R. (Roy), 1936- Main title: Marine geochemistry / Roy Chester and Tim Jickells. Edition: 3rd ed. Published/Created: Hoboken, NJ: John Wiley & Sons, c2012. Description: vii, 411 p., [12] p. of plates: ill. (some col.), maps (some col.); 26 cm. ISBN: 9781118349076 (cloth) 9781405187343 (pbk.) LC classification: GC111.2 .C47 2012 Related names: Jickells, T. D. (Tim D.) Subjects: Chemical oceanography. Marine sediments. Geochemistry. Notes: Includes bibliographical references and index.

Marine tephrochronology LCCN: 2014451556 Main title: Marine tephrochronology / edited by W.E.N. Austin, P.M. Abbott, S.M. Davies, N.J.G. Pearce and S. Wastegård. Published/Produced: London: The Geological Society, 2014. Description: 213 pages: illustrations (some color), maps (some color); 26 cm. ISBN: 9781862396418 1862396418 LC classification: QE527.56 .M37 2014 Related names: Austin, W. E. N. (William E. N.), editor. Abbott, P. M. (Peter M.), editor. Davies, S. M. (Siwan M.), editor. Pearce, N. J. G., editor. Wastegard, S., (Stefan), editor. Abstract: This Special Publication includes articles presenting recent advances in marine tephrochronological studies and outlines innovative techniques in geochemical fingerprinting, stratigraphy and the understanding of depositional processes. It represents a significant resource for the palaeoceanographic community at a time when marine tephrochronology is being more widely recognized. It will also serve as a valuable reference to a much wider community of Earth scientists, climate scientists and archaeologists, particularly in highlighting the role of tephra studies in stratigraphy and regional/extra-regional correlations, as well as in tracing the long-term history of regional and global volcanism in the deep-sea archive. Contents: Marine tephrochronology: an introduction to tracing time in the ocean / W.E.N. Austin, P.M. Abbott, S. Davies, N.J.G. Pearce and S. Wastegård -- Marine tephrochronology: a personal perspective / D.J. Lowe -- Preparation of micro- and crypto-tephras for quantitative microbeam analysis / M. Hall and C. Hayward -- Microbeam methods for the analysis of glass in fine-grained tephra deposits: a SMART perspective on current and future trends / N.J.G. Pearce, P.M. Abbott, and C. Martin-Jones -- Physical characteristics of tephra layers in the deep sea realm: the Campanian Ignimbrite eruption / S.L. Engwell, R.S.J. Sparks and S. Carey -- Identification of a MIS6 age (c. 180 ka) Icelandic tephra within NE Atlantic sediments: a new potential chronostratigraphic marker / F.D. Hibbert, S. Wastegård, R/ Gwynn and W.E.N. Austin -- Marine Ash Zone IV: a new MIS 3 ash zone on the Faroe Islands margin / S. Wastegård and T.L. Rasmussen -- North Atlantic marine radiocarbon reservoir ages through Heinrich event H4: a new method for marine age model construction / J. Olsen, T.L. Rasmussen and P.J. Reimer -- Last millennium dispersal of air-fall tephra and ocean-rafted pumice towards the north Icelandic shelf and the Nordic seas / G. Larsen, J. Eiríksson and E.R. Gudmundsdóttir -- Iceberg-rafted tephra as a potential tool for the reconstruction of ice-sheet processes and ocean surface circulation in the glacial North Atlantic / M. Kuhsk, W.E.N. Austin, P.M. Abbott and D.A. Hodell -- Holocene tephra from Iceland and Alaska in SE Greenleaf shelf sediments / A. Jennings, T. Thordarson, K. Zalzal, J. Stoner, C. Hayward, Á. Geirsdóttir and G. Miller -- Quantifying bioturbation of a simulated ash fall event / J.A. Todd, W.E.N. Austin and P.M. Abbott. Subjects: Tephrochronology. Marine sediments. Submarine geology. Paleoceanography. Volcanic ash, tuff, etc. Notes: Includes bibliographical references and index.

Series: Geological Society special publication, 0305-8719; no. 398 Geological Society special publication; no. 398.

Organic compounds in soils, sediments & sludges: analysis and determination LCCN: 2012036581 Main title: Organic compounds in soils, sediments & sludges: analysis and determination / [editor], T. Roy Crompton, Retired, UK Rivers Authority, UK. Published/Produced: Boca Raton: CRC Press, Taylor & Francis Group, [2013] Description: xii, 263 pages: illustrations; 26 cm ISBN: 9780415644273 (Hbk: alk. paper) LC classification: S592.6.O73 O74 2013 Related names: Crompton, T. R. (Thomas Roy) Subjects: Soils--Organic compound content. Marine sediments. Soil pollution--Measurement. Sewage sludge. Organic water pollutants--Measurement. Notes: "A Balkema Book." Includes bibliographical references and index.

Principles of tidal sedimentology LCCN: 2011939475 Main title: Principles of tidal sedimentology / Richard A. Davis, Jr., Robert W. Dalrymple, editors. Published/Produced: Dordrecht; New York: Springer, [2012] Description: xv, 621 pages: illustrations (some color), maps; 27 cm ISBN: 9789400701229 (hbk) 9400701225 (hbk) 9400701233 (e-book) 9789400701236 (e-book) LC classification: QE471 .P748 2012 Related names: Davis, Richard A., Jr., 1937- Dalrymple, Robert W. (Robert Walker) Subjects: Sedimentology. Marine sediments. Notes: Includes bibliographical references and index.

Reconstructing Earth's climate history: inquiry-based exercises for lab and class LCCN: 2011039577 Main title: Reconstructing Earth's climate history: inquiry-based exercises for lab and class / Kristen St John... [et al.]. Published/Created: Hoboken, N.J.: Wiley, 2012. Description: xlii, 485 p.: ill. (some col.), maps (some col.); 29 cm. Links: Cover image http://catalogimages.wiley.com/images/db/jimages/9780470658055.jpg ISBN: 9780470658055 (hbk.) 9781118232941 (pbk.) LC classification: QC884 .R428 2012 Related names: St. John, Kristen. Summary: "The context for understanding global climate change today lies in the records of Earth's past. This is demonstrated by decades of paleoclimate research by scientists in organizations such as the Integrated Ocean Drilling Program (IODP), the Antarctic Geological Drilling Program (ANDRILL), and many others. The purpose of this book is to put key data and published case studies of past climate change at your fingertips, so that you can experience the nature of paleoclimate reconstruction. Using foundational geologic concepts you will explore a wide variety of topics in this book, including: marine sediments, age determination, stable isotope paleoclimate proxies, Cenozoic climate change, climate cycles, polar climates, and abrupt warming and cooling events. You will evaluate published scientific data, practice developing and testing hypotheses, and infer the broader implications of scientific results. It is our philosophy that addressing how we know is as important as addressing what we know about past climate change. Making climate change science accessible is the goal of this book. Readership: earth science students at a variety of levels studying paleoclimatology, oceanography, Quaternary science, or earth-system science"-- Provided by publisher. "This project integrates scientific ocean drilling data and research (DSDP-ODP-IODPANDRILL) with education"-- Provided by publisher. Contents: Machine generated contents note: The Authors Acknowledgments Book Introduction for Students and Instructors Geologic Timescale Chapter 1. Introduction to Paleoclimate Records. Part 1.1. Archives and Proxies Part 1.2. Owens Lake - An Introductory Case Study of Paleoclimate Reconstruction Part 1.3. Coring Glacial Ice and Seafloor Sediments Chapter 2. Seafloor Sediments. Part 2.1. Sediment Predictions Part 2.2. Core Observations and Descriptions Part 2.3. Sediment Composition Part 2.4. Geographic Distribution and Interpretation Chapter 3. Microfossils and Biostratigraphy. Part 3.1. What are Microfossils? Why are they Important in Climate Change Science? Part 3.2. Microfossils in Deep-sea Sediments Part 3.3. Application of Microfossil First and Last Occurrences Part 3.4. Using Microfossil Datums to Calculate Rates Part 3.5. How Reliable are Microfossil Datums? Chapter 4. Paleomagnetism and Magnetostratigraphy. Part 4.1. Earth's Magnetic Field Today and the Paleomagnetic Record of Deep-Sea Sediments Part 4.2. Paleomagnetism in Ocean Crust Part 4.3. Using Paleomagnetism to Test the Seafloor Spreading Hypothesis Part 4.4. The Geomagnetic Polarity Time Scale Chapter 5. CO2 as a Climate Regulator during the Phanerozoic and Today. Part 5.1. The Short Term Global Carbon Cycle Part 5.2. CO2 and Temperature Part 5.3. Recent Changes in CO2 Part 5.4. The Long-term Global Carbon Cycle, CO2, and Phanerozoic Climate History Chapter 6. The Benthic Foraminiferal Oxygen Isotope Record of Cenozoic Climate Change. Part 6.1. Introduction Part 6.2. Stable Isotope Geochemistry Part 6.3. A Biogeochemical Proxy Part 6.4. Patterns, Trends and Implications for Cenozoic Climate Chapter 7. Scientific Drilling in the Arctic Ocean: A Lesson on the Nature of Science. Part 7.1. Climate Models and Regional Climate Change Part 7.2. Arctic Drilling Challenges and Solutions Part 7.3. The Need for Scientific Drilling Part 7.4. Results of the Arctic Drilling Expedition Chapter 8. Climate Cycles. Part 8.1. Patterns and Periodicities Part 8.2. Orbital Metronome Part 8.3. A Break in the Pattern Chapter 9. The Paleocene Eocene Thermal Maximum (PETM) Event.

Part 9.1. The Cenozoic [delta]13C Record and an Important Discovery Part 9.2. Global Consequences of the PETM Part 9.3. Bad Gas: Is Methane to Blame? Part 9.4. How fast? How long? Part 9.5. Global Warming Today and Lessons from the PETM Chapter 10. Glaciation of Antarctica: The Oi1 Event. Part 10.1. Initial Evidence Part 10.2. Evidence for Global Change Part 10.3. Mountain Building, Weathering, CO2 and Climate Part 10.4. Legacy of the Oi1 Event: The Development of the Psychrosphere Chapter 11. Antarctica and Neogene Global Climate Change. Part 11.1. What do we Think we Know about the History of Antarctic Climate? Part 11.2. What is Antarctica's Geographic & Geologic Context? Part 11.3. Selecting Drillsites to Best Answer our Questions Chapter 12. Interpreting Antarctic Sediment Cores: A Record of Dynamic Neogene Climate. Part 12.1. What Sediment Facies are Common on the Antarctic Margin? Part 12.2. ANDRILL 1-B The BIG Picture Part 12.3. Pliocene Sedimentary Patterns in the ANDRILL 1-B Core Ch. 13. Pliocene Warmth: Are We Seeing Our Future? Part 13.1. The last 5 million years Part 13.2. Sea Level Past, Present, and Future Chapter 14. Northern Hemisphere Glaciation. Part 14.1. Concepts & Predictions Part 14.2. What is the Evidence? Part 14.3. What Caused It? Index. Subjects: Paleoclimatology. Climatic changes--Observations. Climatic changes--History. SCIENCE / Earth Sciences / Meteorology & Climatology. Notes: Includes bibliographical references and index.

Submarine mass movements and their consequences: 5th international symposium LCCN: 2011939086 Main title: Submarine mass movements and their consequences: 5th international symposium / Yasuhiro Yamada ... [et al.], editors. Published/Created: Dordrecht; New York: Springer, c2012. Description: xxxi, 769 p.: ill. (some col.), maps (some col.); 24 cm. ISBN: 9789400721616 (hbk.: alk. paper) 9400721617 (hbk.: alk. paper) 9789048130702 9048130700 LC classification: QE598 .S83 2012 GC Related names: Yamada, Yasuhiro. Subjects: Mass-wasting--Congresses. Submarine topography--Congresses. Landslide hazard analysis--Congresses. Marine sediments--Congresses. Tsunamis--Congresses. Submarine geology. Conference proceedings. Massenbewegung (Geomorphologie) Meeresboden. Submarine Gleitung. Form/Genre: Austin (Texas, 2009) Kongress-Austin (Texas)-2009. Kongress. Notes: Sixty-five papers from an international symposium held in Kyoto, Japan, in 2011. Includes bibliographical references and indexes. Series: Advances in natural and technological hazards research; v. 31 Advances in natural and technological hazards research; v. 31.

INDEX

A

B

C

D

E

F

G

H

I

J

K

L

M

N

O

P

Q

R

S

T

U

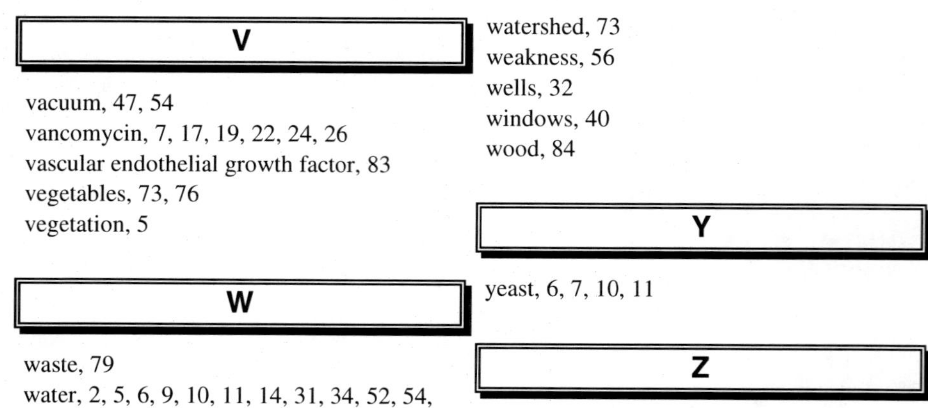

V

W

Y

Z